Process Mining in Action

Lars Reinkemeyer
Editor

Process Mining in Action

Principles, Use Cases and Outlook

 Springer

Editor
Lars Reinkemeyer
Department of Computer Science
University of California, Santa Barbara
Santa Barbara, California, USA

ISBN 978-3-030-40174-0   ISBN 978-3-030-40172-6   (eBook)
https://doi.org/10.1007/978-3-030-40172-6

This Springer imprint is published by the registered company Springer Nature Switzerland AG.
The registered company address is: Gewerbestrasse 11, 6330 Cham, Switzerland

Dedicated to my Family

Foreword

The world is becoming increasingly complex—a mantra heard every day—so it is all the more important to bear it in mind. Because it doesn't just affect everyday life, but also the markets in which companies operate. New players constantly appear and change the game, both from a technical point of view and in terms of the competitive situation. An example of this is the platform economy. The digital transformation has long since arrived in the Business Services world: organizations that want to be successful here need to continuously strive for innovation in order to better handle the complexity of their processes, to improve operational efficiency and excellence and to stay ahead of the game.

Especially in organizations with a huge variety of processes, there is a great deal of complexity which forces to find new and innovative ways to handle these complications and gain more transparency. This is the reason why Siemens started as early as 2011 to exploit and thus become an early adopter of a technology that takes an innovative path in simplifying complexity: Process Mining. It essentially creates transparency on business processes and improvement measures can be derived from it.

Process Mining has evolved as a particularly exciting technical innovation, as it allows a holistic understanding of even the most complex processes and is a good example of how academic research can support established industries and drive innovation in practice. By implementing Process Mining, the data that numerous processes produce during daily work is converted into insights, thus creating a seamless end-to-end experience and transparency.

In the last couple of years, Process Mining has been established in many parts of the Siemens organization and has become a commonly adopted technology which supports operational excellence in Purchase-to-Pay and Order-to-Cash processes. For example, Process Mining drives transparency in all Order-to-Cash processes from incoming orders to payment receipts and helps to quickly identify improvement and automation levers. In Purchase-to-Pay processes, from sending orders to goods receipt and paying invoices, it enables transparency and again helps to quickly identify automation levers, which in turn have directly led to a reduction in operational inefficiency. In this way, a high amount of process variants can be reduced and the remaining once be optimized, digitalization potential identified, and the most favorable payment terms taken.

Process Mining has therefore become a reliable and indispensable single source of truth and provides employees and management, as well as customers, with the same objective insights. This transparency is of utter importance as the future will see technology progress at an increasing pace, opening up the door for new symbioses as already seen by combining Robotic Process Automation and Artificial Intelligence, for example. Process Mining can be used as a fundamental enabler as its transparency provides guidance for automation potentials and Artificial Intelligence use cases. Thus, Process Mining, in addition to Robotics Process Automation and Artificial Intelligence, is a building block on the way to a new way of work that empowers employees with digital tools and ultimately shapes the digital organization of the future.

This book, written by practitioners for practitioners, provides an excellent overview on Process Mining principles and a broad variety of use cases. Each practitioner shares individual experiences, and the multitude of outlooks from practitioners and academics will foster future development and innovations and increase business value, thus, leaving Process Mining as an academic and innovative field that could well be one of the key technologies driving the Digital Transformation.

Enjoy reading!

Siemens AG   Hannes Apitzsch
Munich, Germany

Foreword

Process Mining in a Volatile World

Today's world is driven by volatility, speed, major steps in technology development, uncertainties in political relationships, and a significant change in environmental protection. These topics are propelling the transformation all over the world. The key reason to these changes is a radical implementation of mature IT technologies within the last 10–20 years. In the mobility industry, the effects are zero emission mobility, autonomous driving, shared business models, and connecting every car to the World Wide Web. These are key words of disruption in the automotive world. Technology drivers for this are: worldwide connection to the internet, big data, artificial intelligence, IoT, blockchain, cloud platforms, additive manufacturing/3d printing, standardization of technologies and APIs, cyber security, robots, and process automation. Digitalization is the combination of all these technologies. Digitalization could be described with two words: speed and flexibility.

In the mobility and automotive industry, this leads to three dimensions:

1. The customer of mobility becomes the focus point of the new business models (CRM).
2. The car gets a part of the internet combined with new powertrain solutions (ACES—autonomous, connected, electrified, shared).
3. The current business models will be disrupted (data-oriented, IoT, and cloud platforms).

Thus, business process models have to be transformed into the next generation. Between 2007 and 2010, Nokia has shown that a world leader in mobile technologies will not be part of the game, if not or too slowly reacting to these changes. The transformation of the business process models will be successful if full process transparency (KPI, real time, documentation) is given, the processes are analyzed end to end, and process organization, IT optimization, and reengineering are based on this process transparency.

The BMW Group IT Strategy 1.0 (2016) and IT-Strategy 2.0 (2019) is adopting disruptive changes within the automotive environment. Speed and flexibility are and will be the main characteristics for the next decade. On the one hand, this means the SW engineering process has to be transformed and the base has to be set up for implementing the new technologies and transforming legacy into the future.

The BMW Group IT Program is called "100% agile." Changing from "waterfall SW engineering" to "100% agile SW engineering" with increasing internal competencies enables the implementation of all these new differentiating IT technologies.

Best practice models were taken from digital leaders such as Spotify, Netflix, Google, Salesforce, Airbnb, Amazon, Tencent, and Alibaba. Moreover, a target picture "agile working model" was developed and implemented to become a driver in the tech mobility industry.

In 2016, a small team of the BMW Group IT made an exploration of the Process Mining technology with two proofs of concept. The capability was evaluated in a 12-month project. In the end of 2017, first potentials were checked in the two areas "worldwide purchase to pay process" and "paint shop process of a BMW car manufacturing plant." In the beginning of 2018, the decision to set up a "Center of Excellence for Process Mining" was made. This CoE (process excellence, tool experience, operational excellence) was the driver for building up a worldwide internal competence network (IT and Data). The Process Mining Network had to implement the technology into the BMW Group processes. Many external partners (Software Provider, implementation partner, research partner/universities) were integrated. BMW is an innovator in "process-oriented organizations" since the end of the 1990s. Six core processes were defined: Idea to Offer (development), Order to Deliver (production), Offer to Order and Deliver to Customer Care (sales), Financial Services, Business Management (e.g., strategy, governance), and Resource Management (e.g., finance, IT, human resource, procurement). In all areas, the Process Mining technology was evaluated and implemented. With a cross-functional team, the areas with highest potential for BMW were prioritized. Process Mining helps BMW to improve services and products successfully (tremendous productivity increase), for instance:

- Purchasing and Finance (improves compliance and automation rates)
- Development (reduces development cycle and increases standards)
- Change Management (improve transparency and product quality)
- Production (reduce throughput times and rework rates)
- Leasing (faster processes)
- User Experience (usage of products)
- Aftersales (optimization of customer touch points)
- IT (optimization of the IT system landscape)

Process Mining could be used not only in classical administration processes but also in production, production control, development, financial services, and sales processes. Therefore, Process Mining fits into all areas, where IT is implemented "end to end" and cross-functional. In 2019, reengineering highlights were processes in car distribution, customs, the car navigation system, and in cost analysis of production. Furthermore, the implementation of Process Mining gets sustainable because of the objectification of process performance all over the company. All in all, the transformation of an organization into a digital organization will only be successful if Process Mining tools and methods are implemented in day-to-day operations.

Important in the Process Mining environment is the how and what of the implementation. This book, with many pragmatic and operative examples, gives a guideline and a framework for the utilization in different branches and organizations. It shows us that this new technology will not be a short-time hype. It will be a foundation technology for digitalization of organizations. Take this book as a basic description of Process Mining for the next years. Get advantages out of these examples.

Munich, Germany   Klaus Straub

Preface

Imagine full transparency regarding operational processes and activities as they actually happen in your organization. Imagine the power of such insight to reduce complexity and expedite digital transformation of internal processes.

Complexity is a key challenge for any organization. Large organizations have to deal with hundreds of thousands of suppliers as well as millions of purchase orders, customer orders, deliveries, and financial transactions. Smaller companies equally face the challenge to understand actual business processes and organizational complexity in order to continuously improve operational efficiency, reduce transactional cost, and compete with an increasing number of digital native contestants.

Process Mining, as a key lever to address the complexity challenge, represents perhaps the most exciting technological innovation since the advent of digital transformation. No other technology is capable to provide similar process transparency, allows a similar understanding of actual process performance and thus fuels digital transformation of internal processes based on objective insights and facts. While the underlying principles of Process Mining—collecting digital traces of actual business processes and visualizing complex process flows to allow a thorough understanding and sustainable improvement—are straightforward and have been extensively discussed in academic publications, the devil for operational implementation and impact lies in the detail.

In the last decade many companies started to use Process Mining in an operational environment and gained substantial experiences. Process Mining principles have been adopted to industrial requirements in order to create economic and ecologic value. An increasing number of use cases have been applied along the whole value chain, with a growing community sharing and promoting the potentials and benefits of Process Mining. This book reflects these developments, with best practice use cases from companies which have successfully deployed Process Mining. Contributors have been invited based on their proven success to establish Process Mining in their respective organization and represent a broad range of different industries, functions, and use cases. The book is written by practitioners for practitioners and independent of Process Mining software vendors or consultants in order to provide an unbiased overview of current usage, technological capabilities, and future potentials. It provides hands-on examples and experiences on how to use Process Mining in different organizational environments for strategic insights, digital transformation, and sustainable operations in order to ultimately generate operational impact and value.

In addition to economic value, Process Mining can also contribute to ecologic benefits. As environmental challenges prevail, it is time for a sustainability revolution, and technological innovations such as Process Mining must play a significant role, e.g., to identify process inefficiencies, reduce energy consumption, and support the reduction of material waste. Respective best practices and potential positive impacts have been flagged out throughout the book wherever applicable.

The book complements the large number of academic publications. Process Mining as an academic field of research has been expanding since the beginning of this millennium. It was "invented" by Wil van der Aalst in the late 1990s; he provided groundbreaking research and established a global academic community. Based on a solid academic foundation, Process Mining has received increasing interest from operational business with many companies already harvesting benefits. The market has seen the rise of new software companies providing innovative solutions and an increasing awareness of the power of this innovative technology, which is becoming a foundation technology for the digitization of organizations.

Commercial usage of Process Mining started earlier this decade, initially with a strong focus on single projects, e.g., for audit with one-time effort, short duration, and limited scope. As the value of Process Mining became more transparent and relevant for digital transformation, it extended across different business functions, such as Procurement, Sales, Finance, and Logistics, and different industries, such as Manufacturing, Automotive, Telecommunication, Healthcare, Insurance, and Aeronautics. The technology is used in single departments for operational improvement and as part of strategic programs to drive digital transformation and increase organizational efficiency. In large organizations like BMW, Siemens, and Telekom, it is used enterprise-wide as a standard tool for digital transformation and identification of process inefficiencies and reduction of waste.

Complementary to corporate applications, which provide a centrally defined standard format used by a large numbers of users, there are communities of individual users applying the technology to unique local use cases. The book describes use cases and business impact along the value chain, from corporate to local applications, developed with functional experts and thus representing state-of-the-art domain know-how.

Besides success stories, the book discloses challenges, learnings, and failures in order to share experiences with the reader and provide guidance on how to avoid pitfalls and assure a successful operational deployment.

Outline of the Book

The book is structured in three parts: Part I provides an introduction to the topic, from fundamental principles to key success factors and an overview of operational use cases. As a wholistic description of Process Mining in a business environment, this part is probably more beneficial for readers not yet very familiar with the topic.

Part II presents 12 use cases written by contributors from multiple functions and industries. As use cases are presented in detail, this part is probably more beneficial for users who are already familiar with the topic. Part III provides an outlook on the future of Process Mining, both from an academic and an operational perspective.

Part I sets the stage by describing the principles and value of Process Mining. Principles such as event logs and process variants are discussed to explain how Process Mining can provide process transparency and create business value. As a practical guide, hands-on experiences on how to get started are shared, including recommendations for how to initiate first successful projects. Three aspects are crucial for the success of any project, defined as the "three Ps" of Process Mining and discussed in one chapter each:

- Purpose: the specific demand or Purpose should be defined by the Process Owner before starting any project to assure that technology will deliver value by providing insights which will be turned into operational actions.
- People: as valuable as a digital tool can be—engaging the right people to turn insights into action is crucial. The right digital mindset is a prerequisite to improve operational efficiency.
- Processtraces: Process Mining is based on event logs and the identification, collection, and customization of digital processtraces is a major effort driver.

Sharing challenges, pitfalls, and failures, Chap. 6 discloses numerous experiences which are discussed to help avoid similar experiences during the reader's journey. Further technologies used for digital transformation, such as Robotic Process Automation (RPA) and Business Process Management (BPM), will be set into relation with Process Mining and the concept of a Digital Twin of an Organization (DTO) will be explained.

The part is concluded with a summary of ten key learnings, reflecting the key messages from the previous chapters.

Part II presents 12 use cases from companies which have successfully applied Process Mining for different purposes, from optimization of customer journeys to manufacturing processes, from improvement of supply chain management to service processes, from strategic reinvention to product management inventions, and from digitization of internal audit to automation of finance processes. All use cases have been written by Process Mining experts, who are independent from any particular software vendor.

Prerequisite for the selection of use cases has been, that the respective project was implemented with tangible benefits. To allow consistent reading, all use cases are presented in a standardized structure, including Abstract, Challenges, Use Cases, Impact, Technology, Lessons Learned, and Outlook. The use cases represent a variety of different industries, functions, operational roles (e.g., Management, Process Owner, Consulting), and software vendors. It allows to learn from practitioners as they describe which challenges they have faced, which use cases have created impact, and which lessons have been learned and provide an outlook on the future of Process Mining.

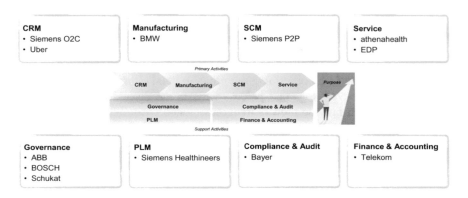

Fig. 1 Use Cases

To visualize the widespread of use cases across different functions, a value chain model has been applied with primary functions (CRM, Manufacturing, SCM, Service) and support activities (Governance, PLM, Compliance & Audit, Finance & Accounting). This model, which is shown in Fig. 1, is similarly used to discuss the purpose of Process Mining in Chap. 3.

Part III gives an outlook on the future of Process Mining, both from an academic and from a business perspective. The academic perspective is contributed by Professor Wil van der Aalst, who "invented" Process Mining. Wil describes the evolution of Process Mining, reveals the inconvenient truth regarding commercial software, discusses the most pressing novel challenges, and concludes with his vision of Business Process Hygiene, which implies that *not* using Process Mining should require a justification—and not the other way around.

The business perspective provides an operational outlook and takes a liberty for prediction and vision, which would not have been appropriate for an academic outlook. Aspects of the prediction include business requirements, technological developments, Artificial Intelligence (AI), strategic workforce planning, and anticipated market developments. As any prediction is difficult—especially about the future—this part reflects the editor's personal view and aims to steer further reflections and discussions for the future evolution of Process Mining.

The book concludes with a vision for a Digital Enabled Organization.

Santa Barbara, California, USA   Lars Reinkemeyer

Acknowledgments

Writing this book became an even more inspirational journey than initially anticipated. The first ideas were born with the positive, constructive, and collaborative vibes in the Process Mining community, thought leaders like Alex, Olly, and many more. As exciting operational use cases became public, it felt worthwhile to promote experiences and provide a stage to the growing number of evangelists. Writing to all major Process Mining vendors received positive feedbacks and exceeded the initial target of 10 cases.

A very special thanks is owed to the contributors, without whom this colorful mosaic would never have been possible. You have done great, not only to drive cases in your company, but to share your operational experiences, adopt the standard structure as a corset for your contribution, and deliver perfectly on the happy path.

The two forewords from Hannes Apitzsch and Klaus Straub are a distinction for Process Mining. Innovations require courageous frontrunners as sponsors, and it is highly appreciated to have you providing a Top Management patronage for this topic.

On the academic side, Wil van der Aalst has not only invented this exciting technology, but stirred discussions with insights into the inconvenient truth and novel challenges. Tom Davenport and Seungjin Whang condensed our inspiring discussions and presentations in concise blurbs. Jianwen Su made the book possible by supporting my scholarship and with interesting discussions in his faculty at UCSB.

At Siemens, there have been—besides Hannes Apitzsch—numerous supporters from the very early days. And it reflects an amazing corporate culture to grant a sabbatical, which was approved by Erik Mohr and Dietmar Mauersberger.

When the first seeds for the project disseminated, my friends gave me the courage to start on this adventure. On the journey, Al Gore extended my horizon with his plea for a sustainability revolution. And my father was of great help not only by proofreading the draft version, but also by extending my horizon with his wealth of experiences.

Last but not least I would like to thank my family—including King Jack—for joining me on this very special adventure.

Contents

Part I Principles and Value of Process Mining

1 Process Mining in a Nutshell 3
Lars Reinkemeyer

2 How to Get Started 11
Lars Reinkemeyer

3 Purpose: Identifying the Right Use Cases 15
Lars Reinkemeyer

4 People: The Human Factor 27
Lars Reinkemeyer

5 Processtraces: Technology 31
Lars Reinkemeyer

6 Challenges, Pitfalls, and Failures 37
Lars Reinkemeyer

7 Process Mining, RPA, BPM, and DTO 41
Lars Reinkemeyer

8 Key Learnings 45
Lars Reinkemeyer

Part II Best Practice Use Cases

**9 Siemens: Driving Global Change with the Digital Fit Rate
in Order2Cash** 49
Gia-Thi Nguyen

**10 Uber: Process Mining to Optimize Customer Experience and
Business Performance** 59
Martin Rowlson

11 BMW: Process Mining @ Production 65
Patrick Lechner

12   **Siemens: Process Mining for Operational Efficiency**
      in Purchase2Pay   75
      Khaled El-Wafi

13   **athenahealth: Process Mining for Service Integrity**
      in Healthcare   97
      Corey Balint, Zach Taylor, and Emily James

14   **EDP Comercial: Sales and Service Digitization**   109
      Ricardo Henriques

15   **ABB: From Mining Processes Towards Driving Processes**   119
      Heymen Jansen

16   **Bosch: Process Mining—A Corporate Consulting Perspective**   129
      Christian Buhrmann

17   **Schukat: Process Mining Enables Schukat Electronic to Reinvent**
      Itself   135
      Georg Schukat

18   **Siemens Healthineers: Process Mining as an Innovation Driver**
      in Product Management   143
      Jutta Reindler

19   **Bayer: Process Mining Supports Digital Transformation**
      in Internal Audit   159
      Arno Boenner

20   **Telekom: Process Mining in Shared Services**   169
      Gerrit Lillig

Part III   **Outlook: Future of Process Mining**

21   **Academic View: Development of the Process Mining Discipline**   181
      Wil van der Aalst

22   **Business View: Towards a Digital Enabled Organization**   197
      Lars Reinkemeyer

About the Editor   207

Abbreviations

ACL	Audit Command Language
AHT	Average Handling Time
AI	Artificial Intelligence
API	Application Programming Interface
A/P	Accounts Payable
A/R	Accounts Receivable
BI	Business Intelligence
BP	Business Process
BPH	Business Process Hygiene
BPM	Business Process Management
BPMN	Business Process Management and Notation
BTO	Build to Order
BVI	Business Volume Indicator
B2B	Business to Business
B2C	Business to Consumer
CoE	Center of Excellence
CT	Computer Tomography
DaaS	Data as a Service
DFG	Direct Flows Graph
DMAIC	Define, Measure, Analyze, Improve, Control
DTO	Digital Twin of an Organization
EDI	Electronic Data Interchange
ERP	Enterprise Resource Planning
ETL	Extract, Transport, Load
EWM	Enterprise Warehouse Management
e2e	end to end
FIFO	First In First Out
GDPR	General Data Protection Regulation
HQ	Headquarter
HR	Human Resources
IoT	Internet of Things
IIoT	Industrial Internet of Things
KPI	Key Performance Indicator

LC	Lead Country
LIFO	Last In First Out
MES	Manufacturing Execution System
ML	Machine Learning
MRT	Magnetic Resonance Tomography
MVP	Minimum Viable Product
NPS	Net Promoter Score
OCR	Optical Character Recognition
OPQ	Opportunities for Perfecting Quality
O2C	Order to Cash
PLM	Product Lifecycle Management
PMI	Process Management Insights
PO	Purchase Order
PoC	Proof of Concept
PPI	Process Performance Indicator
PR	Purchase Requisition
PTP	Procure to Pay
P2P	Purchase to Pay
RCA	Root Cause Analysis
RDS	Relational Database Service
RoI	Return on Investment
RPA	Robotic Process Automation
RPD	Robotic Process Discovery
RPM	Robotic Process Management
SCM	Supply Chain Management
SDI	Smart Data Integration
SLA	Service Level Agreement
SLT	SAP Landscape Transformation
SQL	Structured Query Language
SRM	Supplier Relationship Management
S2P	Source to Pay
UCSB	University of California, Santa Barbara
UI	User Interface

Part I

Principles and Value of Process Mining

How do you drive digital transformation without thorough understanding of current processes and complexity drivers? Can you afford to manage the necessary transformations in your organization without using technical possibilities, which allow you to achieve full transparency about actual processes and complexity drivers?

Process Mining is one of the most innovative and maybe the most exciting digital tools supporting companies on their journey towards digital transformation. It provides wholistic insights into actual processes and complexities, thus allowing to identify inefficiencies and effort driver. Imagine a digital tool which allows you to visualize and understand any business process in your organization—from every single process activity to an aggregated global view—based on the actual event logs of activities. Imagine full transparency into process complexities and perfect insights, allowing you to drive transformation and initiate actions such as process redesign, workflow optimization, batch processing, and activity automation with bots, thus increasing the efficiency of internal business processes.

Key Learning #1: Transparency Is a Prerequisite for Digital Transformation

The following chapter describes the fundamental principles of Process Mining and how it is used to create value. The value of Process Mining has been documented in numerous use cases, e.g., due to substitution of process inefficiencies and improvements of sales processes, yielding in savings of millions of euro and further value which has not been quantified. Process Mining can support the transition towards a more efficient and sustainable economy, e.g., in the field of supply chain, by allowing to increase transport efficiency, avoidance of empty transports, optimization of inventories, and reduction of waste. Insights are also used for an optimized transport modal change, thus leading not only to economic but also ecologic value.

Process Mining in a Nutshell

1

Lars Reinkemeyer

Abstract

Fundamentals such as event logs, cases, activities, and process variants are explained. Concrete examples show how Process Mining can be used for business transparency and value. Allowing full transparency based on event logs, the implications of this important change—away from perception based towards a fact-based process management—are discussed. The metaphor of an MRT is used to explain possibilities, benefits, and limitations of Process Mining.

"Process Mining is a process management technique that allows for the analysis of business processes based on event logs." This definition by Wikipedia embraces the unique approach of Process Mining to allow the analysis of any business process based on digital traces captured in event logs. An event log is a collection of events which have taken place in order to perform a business process. Each event refers to a specific activity that took place at a certain moment in time and can be assigned to a unique case. An event log consists—as a minimum requirement—of a Case ID as a numeric identifier, an Activity as a specification of which activity has taken place and a Timestamp for the precise time of every action taken (see Fig. 1.1).

Further attributes can be added to provide further information about the specific activities. In a corporate environment, event logs are digital traces which are stored for any business activity in databases such as ERP, CRM, SRM, MES, or PLM systems. Each customer offer, order, invoice, etc., is processed in a database where it

L. Reinkemeyer (✉)
University of California, Santa Barbara, Santa Barbara, CA, USA
e-mail: reinkemeyer@ucsb.edu

© Springer Nature Switzerland AG 2020
L. Reinkemeyer (ed.), *Process Mining in Action*,
https://doi.org/10.1007/978-3-030-40172-6_1

Fig. 1.1 Event log

Case ID	Activity	Timestamp
1	Create ticket	January 2, 3:15 PM
1	Screen ticket	January 2, 3:32 PM
1	Repair - simple	January 10, 9:45 AM
1	Close ticket	January 10, 11:34 AM
2	Create ticket	January 2, 4:04 PM
2	Screen ticket	January 2, 4:05 PM
2	Repair - complex	January 3, 1:38 PM
2	Close ticket	January 4, 9:23 AM
3	...	

leaves digital traces. Detailed descriptions of procedures on how to build valuable event logs from data that is stored in relational databases are publicly available.[1]

In order to "mine" a process, the digital traces are identified, extracted, and visualized in a form which reflects the actual process flow, thus providing transparency regarding the sequence of activities as they have actually taken place. Processing time and sequence of the events provides a complete picture of each case, allowing to trace process flows, understand delays, separate loops, and identify complexity drivers.

Figure 1.2 shows the principle of Process Mining to visualize simple and complex process flows in the form of different variants: the left picture presents a simple, standardized process flow. The right picture presents a more complex process flow with different variants, reflecting multiple options for how the process can be performed.

While people tend to design and think in the form of simple process flows (left image), reality tends to be more complex with multiple variants (right image). In this respect, it is commonly differentiated between "To-Be" and "As-Is" processes:

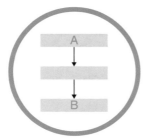 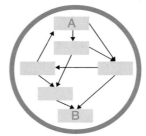

Fig. 1.2 Principle of process variants

---

[1]Example: http://businessinformatics.be/2016/12/20/from-relational-database-to-valuable-event-logs-for-process-mining-a-procedure/

Fig. 1.3 Most common
process variant

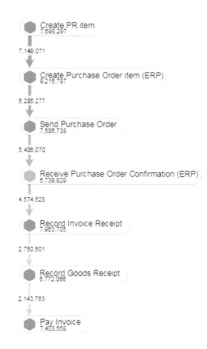

To-Be presents the ideal, perfect process flow without friction, as it is typically designed in theory. As-Is on the other hand presents the actual process flow with all deviations and complexities, which occur in real-life operational processes.

Process Mining allows to visualize any process complexity with the option to drill down from the most simple variant to more complex visualizations of multiple process variants until the ultimate visualization of all process variants, thus reflecting a full picture of activities as they were actually performed.

Figure 1.3 shows an example of the most simple = most common variant of a Purchase to Pay (P2P) process, from "Create PR item" (create Purchase Requisition item) to "Pay Invoice." In this most simple variant, the most common activities are shown in their actual sequence. The numbers under the activities indicate how often the activity has occurred (e.g., "Create PR item" occurred 7,698,297 times), while the numbers on the connection arrows indicate how often an activity was directly followed by the subsequent activity (e.g., in 7,149,071 cases the activity "Create PR item" was directly followed by the activity "Create Purchase Order (PO) item"). As the activity "Create PR item" occurred more often than indicated in the connection to the next activity "Create PO Item," it is an indication that there were other subsequent activities. The activity "Create PO item" occurred 9,216,797 times, indicating that many of these activities did not have the activity "Create PR item" as a direct predecessor, but some other activity, which is not shown in this most common variant.

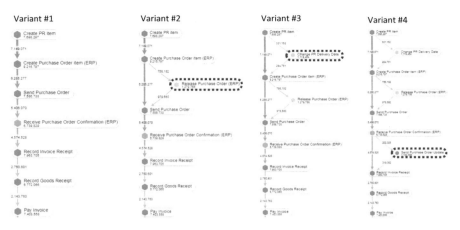

Fig. 1.4 Process variants, example for a P2P process

Process Mining allows to drill into any process step by step, showing further variants with further activities. Figure 1.4 shows the four most common process variants of a P2P process, each disclosing one additional activity. Variant #1 has been described above. Variant #2 shows the second most common variant, where the additional activity "Release PO" has been logged with 1,279,756 cases. Variant #3 shows the third most common variant, where the additional activity "Change PR Delivery Date" has been logged with 1,119,484 cases. And Variant #4 shows the fourth most common variant were the additional activity "Send PO update" has been logged with 1,100,566 cases.

As an alternative to the number of subsequent activities (as shown on the connection arrows) the time required between two activities can be displayed. Calculated as average or median time, this provides insight into which process steps have long durations and might be the reason for delays or bottlenecks.

> *Key Learning #2: Process Mining Allows Full Transparency Based on Event Logs*

Based on these principles, Process Mining allows to drill down step by step into all actual process variants until all cases are included, thus providing an exact picture of the full complexity of the actual processes. Figure 1.5 samples in an excerpt the full complexity of an order to cash (O2C) process, which has several hundred thousand different process variants. The insight allows full transparency to how customer orders are actually processed throughout the organization.

The fun part starts with the surprising insights, which can be gained for specific use cases and which—in almost all cases—reveals numerous surprises even for experienced process experts. The excerpt of O2C gives an idea about the insight into

Fig. 1.5 Excerpt of process complexity, example of an O2C process

complex global processes, thus allowing the identification and a thorough under-standing of operational (in)efficiencies. Based on these insights, dedicated measures for process improvements can be defined and deployed and impact can be continu-ously measured. Dedicated use cases allow to filter those activities and variants which are of special interest, such as delay, rework, or effort driver. Examples of surprising process variants, which were only discovered with Process Mining, have been presented by multiple companies and include the following samples:

- In procurement, more than ten approval steps for the purchase of a single computer mouse could be identified. The different approvers were not aware how many other colleagues also had to approve the same PO.
- In accounting, more than six activity loops were necessary to clarify single payment blocks due to insufficient documentation and authorization rights.
- In accounting, Process Mining allowed to identify cases, where a daily batch process deleted entries which had been previously keyed in manually by human users.
- In logistics, the reason for late deliveries could be discovered by understanding the end-to-end (e2e) throughput time. Insights allowed to identify that selected freight forwarders didn't operate on the date which was requested for delivery.

Process Mining allows a fundamental change in how to work on process analytics and optimization: a traditional project requires extensive observation and manual documentation of actual process steps. Depending on the invested time and effort, a more or less representative number of observations can be collected and prepared, e.g., in Excel or PowerPoint. Based on extrapolations, some assumptions can be deduced in order to estimate actual process flows. However, this approach has obvious limitations, as it only allows to draw a rough picture of the actual reality which is open to biased interpretation and neglects the technical possibility to get full, fact-based transparency.

With increasing data processing power, access to large amounts of event logs, and performant visualization tools, these limitations have been lifted. Today, insights are possible based on every single activity, thus providing a comprehensive picture of all processes, even with millions of single activities. Visualizations allow aggregated views as well as detailed views, thus providing a concise picture of which can be used throughout the whole organization, by top management to operational employees. As a single source of truth, the provided transparency is consistent across all organizational levels.

Process Mining changes the way work is done, as it allows for unanimous objectivity: while previous process analytics projects had to rely on an excerpt of reality and were open for interpretation, today's insight and transparency are based on the entity of event logs and thus show indisputable, actual process flows. In daily business, this represents an important change, away from a perception-based towards a fact-based discussion. The adoption of this new kind of insights typically follows an evolution: as a first reaction, process experts tend to challenge the results and try to find examples for faults. And experience shows that the forgiving for human error is much higher than the forgiving for computer error. Once the "credibility" has been accepted, the focus shifts towards understanding the results and comparing these with current reports and Key Performance Indicators (KPIs) (Fig. 1.6).

Experience shows that once the Process Mining results are commonly accepted, less time is spent on discussing *what* is happening, but the focus of discussion shifts towards *why* things are happening. The analysis of reasons for delays and loops is the first step towards thorough process understanding and improvement. Based on the in-depth insights, actions are defined, deployed, and continuously tracked.

Process Mining allows to drive process understanding and process optimization deeper into the fabric of the organization, as it provides valuable insights on all organizational levels and enables all players to contribute. In many companies the process responsibility is formally assigned to a central process owner, who is in charge of process design, improvement, and efficiency. But operational experts, who execute process activities with specific domain know-how and "live" the process every day, have a significant impact on process efficiency and must support to drive the change. The head of procurement might be the process owner for the P2P

Fig. 1.6 Perception

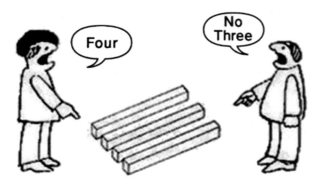

process, but the operational procurement team executes activities and can make immediate improvements. As Process Mining provides insights into all organizational levels, it fosters a decentralization of process improvements and a new collaborative approach as process understanding is faster absorbed throughout the organization, with employees on all levels being enabled to contribute to sustainable improvements. And the unbiased data foundation can facilitate a better collaboration across organizational silos, as the EDP use case in Chap. 15 presents. For BMW (use case in Chap. 11), Process Mining has even become a game changer, as it has substituted the traditional process analytics approach.

The key value of process insights is not that it allows to strive for perfect standardization of processes or add more activities and make flows more complex. On the contrary, insights allow to focus on those variants and activities which are effort drivers and of no value. "Perfection is achieved, not when there is nothing more to add, but when there is nothing left to take away" (Antoine de Saint-Exupéry). In an operational project, the majority of process variants and activities are usually of little interest, as there is "nothing left to take away," which means that all activities and variants are necessary. Focus must be on identifying the exceptions, which cause delay and additional effort with unnecessary activities and variants. With the new insight the process expert is enabled to drill down and decide which process activities and variants should be avoided or substituted on the strive towards process perfection. Perfection can also be defined as standardization, as the example of Uber in Chap. 10 shows, where Process Mining has been used for global process standardization.

The analogy with X-Ray or Magnetic Resonance Tomography (MRT) pictures can be used to explain the principle of Process Mining, as shown in Fig. 1.7: while MRT as a medical tool allows to screen the human body, Process Mining as a digital tool allows to screen business processes. While providing wholistic pictures of bodies/processes, these provide a quick diagnosis of conspicuous issues, detailed analysis, and targeted remediation. The technology (aka MRT machine/Process

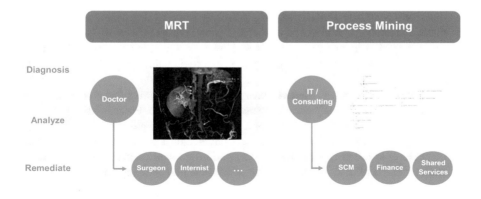

Fig. 1.7 MRT versus process mining

Mining application) provides insight which can be used by experts (aka medical therapist/process expert or consultant) to support the owner (aka patient/process owner).

Throughout the years, this metaphor has been sharpened with several further aspects:

- *Purpose*: MRT technology allows amazing insights, but it needs to be applied for a predefined Purpose which has to be agreed with the patient. Process Mining allows amazing process insights, but it needs to be properly adapted jointly with the process owner for a predefined Purpose. Use cases are a common vehicle to define Purpose jointly with the process owner prior to applying the tool.
- *Therapy*: MRT makes no sense if the patient is not prepared to undergo a therapy once issues have been identified. Process Mining makes no sense if the process owner is not prepared to adopt the results and set up a project to remediate issues which have been identified.
- *Usability*: Using an MRT machine requires training and education. Using a Process Mining tool requires training and education, i.e., to bridge the gap between technology and business to build meaningful reports which are easy to use and valuable.
- *Investment*: Procurement and operations of the MRT equipment is a significant investment. Procurement and operations of Process Mining software usually also come as a significant investment, unless open-source software is used. In any case, investments for deployment and training of this digital tool should be considered as they typically will account for at least 50% of overall cost.
- *Future readiness*: Investment in MRT technology should be future proof, the system should be ready to be upgraded with technological innovation such as Artificial Intelligence (AI). The Process Mining application provider should provide a clear vision on how the tool will be further developed in respect to innovation, scalability, and keeping it future proof.
- *Vision versus reality*: MRT technology promises exciting use cases with AI, e.g., with the automatic detection of conspicuous features and proposal of appropriate therapies. Similarly Process Mining promises exciting use cases, such as automatic detection of digital traces, AI for proactive user support, and Machine Learning (ML) to optimize processes. However, operational experience shows that many of those promises still must prove their value in the face of domain experts.
- *Compulsory Screening*: Medical prevention requires compulsory screening for prevention, or to identify serious issues at an early stage. Process Mining should similarly be applied as a compulsory, preventative screening in order to assure "hygienic" efficient processes. Given today's data and technology, a process owner should rather explain why Process Mining is *not* used for process optimization instead of discussing the value of process transparency and screening.

How to Get Started

2

Lars Reinkemeyer

Abstract

One of the most common questions raised during discussions, presentations, and initiation of projects is the question "how to start a successful project?" Experience and market research show that many Process Mining projects fail. Exaggerated promises and unrealistic expectations, unspecific targets, reluctant teams, and insufficient digital traces can be some reasons for failure. While there is no silver bullet, experience shows that—besides the three factors Purpose, People, and Processtraces, which will be explained in the following chapters—ten general aspects, including quick start, expectation management, and which process to start with, are crucial.

One of the most common questions raised during discussions, presentations, and initiation of projects is the question "how to start a successful project?" Experience and market research show that many Process Mining projects fail. Exaggerated promises and unrealistic expectations, unspecific targets, reluctant teams, and insufficient digital traces can be some reasons for failure. While there is no silver bullet, experience shows that the following three factors—labeled as the "3Ps" of Process Mining—are crucial:

1. Purpose: what is the Purpose for the usage of this digital tool?
2. People: involving the right drivers with a digital mindset for change.
3. Processtraces: availability and reliability of event logs.

L. Reinkemeyer (✉)
University of California, Santa Barbara, Santa Barbara, CA, USA
e-mail: reinkemeyer@ucsb.edu

© Springer Nature Switzerland AG 2020
L. Reinkemeyer (ed.), *Process Mining in Action*,
https://doi.org/10.1007/978-3-030-40172-6_2

Starting a project requires a clearly defined business Purpose and involvement of highly motivated People who can rely on complete, easily accessible Processtraces. All three factors must be in complementary sync. A combination of clear Purpose and motivated People will not be able to succeed if sufficient Processtraces are not available. A combination of clear Purpose and perfect Processtraces will not succeed, if People in the team are reluctant to drive digital transformation. Motivated People and excellent Processtraces will not succeed if there is no clearly defined and achievable Purpose. The three factors have to complement each other to make a project successful.

Key Learning #3: Purpose, People, and Processtraces Are Essential ("3Ps")

To get started with the right project in an operational environment, sufficient time should be invested to identify the best constellation, where all three factors are in good sync. Preparation should include the selection of the most suitable Purpose, identification of the right People, and availability check for sufficient Processtraces. Due to the importance of each of these three factors, they will be discussed in detail in the following three chapters.

In addition to these 3Ps, experience taught the following top ten general aspects and strategies, which are recommended for starting a project:

- *Quick start*: Begin with a simple pilot which allows for a quick start and quick results. Sprinting for a first Minimum Viable Product (MVP), which shows initial value, is important to keep customers and teams excited, create awareness, and inspire other colleagues. Successful launches had first sprints of 2–3 weeks, accompanied by several consecutive demos to show quick results, in an increasing complex format. Progress motivated the team and management to proceed. Starting small, some of these applications have developed into *de facto* corporate standards with onsite teams supported by near- and offshore teams to operate applications, which are used by hundreds of people around the world.
- *Simplicity*: Start with a small, short-term pilot where complexity is reduced to the max. Simplicity includes a clearly defined and focused Purpose (e.g., payment term transparency), a few selected People who are passionate to drive digital change in a first unit (e.g., one department or country only), and easy access to data (e.g., one single data source and structured data). Complexity will grow with time, as the Purpose will be enhanced and more People will get involved.

Key Learning #4: Start with Simplicity to Fight Complexity

- *Expectation management*: Process Mining is a popular topic and some promotions from software vendors and consultants read like excerpts from a brave new world. These promises bear the risk of exaggerated expectations, which can't be fulfilled. In order to manage high expectations, it is recommended to be transparent not only about opportunities, but also about risks and efforts. At the same time it is recommended not to start with the latest technological innovation, but rather with proven technology.

- *Which process to start with?* While Process Mining is in principle applicable to any business process, experience shows that the approach is most easily adopted for homogeneous horizontal support processes. P2P and O2C are classical starting points, as they usually represent relatively standardized processes where saving potentials can be realized with standard use cases such as Rework or Payment Term analysis. Purpose has been proven in multiple use cases, People are typically experienced and digitally minded, and Processstraces easily available. Visualizing logistical and customer journey processes typically is more challenging due to dispersed data sources and heterogenous processes. Manufacturing and Human Resources (HR) typically turn out to be highly challenging due to individualization of manufacturing processes and data privacy requirements for handling HR data.
- *Scope of Impact*: Be clear about the ultimate scope of impact which you wish to achieve long term. While focus on one single and simple process is recommended for a quick start, a successful initial project will lead to an increasing scope. The operational use cases described in the following chapters present examples for different scope, from simple payment term use cases to digital transformation of a corporate process such as O2C up to full corporate digital transformation as in the Schukat use case.
- *Project Team*: The team setup will require members from functional departments (e.g., procurement or sales), IT, and possibly external consultants. Business proximity and domain knowledge should be equally represented as technical understanding for Processstraces and data handling. Assure a project setup with a clear structure and meaningful, achievable targets. The project sponsor should be empowered and determined to adopt recommendation for sustainable improvements. During the project execution, assure smooth and trustful collaboration between functional process owner and the IT department, including joint subtargets and clear responsibilities. Involve the right change agents and recognize also small successes, as they can make a difference.
- *User experience*: For a broad acceptance, the user should have fun working on the application and enjoy an intuitive user interface which provides quick responses. Sharing positive user experiences of some lead users will help to address reluctance of other users, who might be more hesitant to adopt new digital tools.
- *Implementation experience*: It might be beneficial to use the experience of external consultants, who have a proven implementation track record and can facilitate a quick deployment. In-house competences should be established in parallel to assure fast scaling and long-term success.
- *Data security, confidentiality, and privacy*: Secure and confidential handling of required data should be assured right from the start. In particular when working with external consultants, be cautious about providing access to data as, e.g., information on suppliers or customers may be classified as confidential. Many

projects have failed as no access to relevant data has been granted by the responsible data owner or by people responsible for the hosting system. Data privacy should be a crucial point for consideration, not only since the introduction of the EU General Data Protection Regulation (GDPR). As individual behavior might be identified based on particular activity traces and timestamps—even if the user has been anonymized—a responsible data management with an early involvement of the workers council is recommended.

- *Budget*: Depending on the approach you choose—from open-source to chargeable software—budget will be a factor to consider. Gartner lists 19 vendors of Process Mining software with a wide range of technology and price.[1] Price models vary with license fees payable per user, process, or usage. License cost will in most cases be a major cost factor, but don't neglect to budget for data discovery and deployment. Depending on the complexity of the project and the Processtraces, the investment for deployment might exceed the investment for license fees.
- *Don't be scared of failure.* Applying new digital tools requires perseverance.

---

[1]Gartner Market Guide for Process Mining, 06/2019.

Purpose: Identifying the Right Use Cases

3

Lars Reinkemeyer

Abstract

Purpose implies a clear understanding of what Process Mining shall be used for, i.e., which use case shall be investigated. Like for any other tool, an idea is to be formulated first: what shall be achieved and how the tool can contribute. The chapter starts with typical questions from functional departments and reflecting challenges from process owners along the value chain. Examples for purpose, which Process Mining can support, are explained with 22 standard use cases.

Purpose implies a clear understanding of what Process Mining shall be used for, i.e., which use case shall be investigated. Like for any other tool, a clear idea is to be formulated: what shall be achieved and how the tool can contribute. Applying the wrong tool for an unsuitable Purpose will fail. To use an analogy: there is no point in using a hammer to tighten a screw. Equally, there is no point in using Process Mining for building a financial reporting, as other tools are much better suited for this purpose.

Purpose may be defined in an increase of operational process efficiency in order to achieve economic and ecological benefits. Purpose may be defined in order to substitute mundane tasks and enable employees to take responsibility for digital transformation based on better process insights. Purpose may be defined to achieve higher customer satisfaction, less rework, faster deliveries, lower transactional cost, less traffic, or less waste. A good Purpose, which is clearly defined and communicated, will lead to higher efficiency as well as to higher employee engagement and satisfaction.

L. Reinkemeyer (✉)
University of California, Santa Barbara, Santa Barbara, CA, USA
e-mail: reinkemeyer@ucsb.edu

© Springer Nature Switzerland AG 2020   15
L. Reinkemeyer (ed.), *Process Mining in Action*,
https://doi.org/10.1007/978-3-030-40172-6_3

To structure the subsequent discussion about Purpose, we will start with typical questions from functional departments, following a standard value chain. These questions are shown in Fig. 3.1 and reflect challenges from process owners and help to understand for what purpose Process Mining has been applied, i.e., where and how Process Mining can help fight complexity, provide new insights, and support digital transformation.

- *CRM*: A driving factor in sales is a swift customer journey and high customer satisfaction. The Purpose of Process Mining might be defined in whether customer orders are processed quickly and efficiently. Delayed orders may be identified, reasons for delays discovered, and customer satisfaction improved.
- *Manufacturing*: Value stream optimization is essential for productivity, throughput, reduction of working capital, and reduction of waste. The purpose here might be to identify bottlenecks in the value stream, optimize inventories, and reduce working capital as well as material waste.
- *SCM*: On-time delivery to customers is crucial for customer satisfaction. A Purpose can be specified to identify reasons for late deliveries and delays in the logistics process from when a customer order is received to when the goods are delivered at the customer's premises.
- *Service*: For monitoring and improving service performance, a purpose may focus on ticket handling and response times, understanding reasons for delay and complexity drivers.
- *Governance*: Understanding corporate governance as a set of processes, customs, policies, and regulations with which the organization works, a Purpose may be defined to drive process automation and assure policy compliance.
- *PLM*: Understanding actual customer usage of a product can support PLM. Purpose can be defined to identify customer processes while using a product in order to improve the usability of that product.
- *Finance and Accounting*: Purpose can be defined to identify deviations from standard payment terms or deviations between payment terms on POs versus PO confirmations.
- *Compliance and Audit*: As compliance and audit require operational insights, a Purpose can be defined to support compliance checking and auditing process efficiency during walkthroughs and for sampling based on wholistic data insights.

Purpose comes first and needs to be clearly articulated in the form of use cases to support answering the questions prioritized by the process experts. Based on a clear understanding of the different Purposes, a range of operational use cases have proven valuable across the different functions, as shown in Fig. 3.2.

The following discussion of these use cases shall give a concise overview of how Process Mining can add value for different functions. A minor adoption has been

Are customer orders processed quick and efficient?

Have we optimized our value stream and working capital?

What causes late deliveries to customers?

Are tickets answered on time?

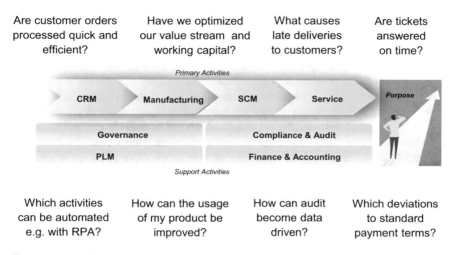

Which activities can be automated e.g. with RPA?

How can the usage of my product be improved?

How can audit become data driven?

Which deviations to standard payment terms?

Fig. 3.1 Purpose/questions

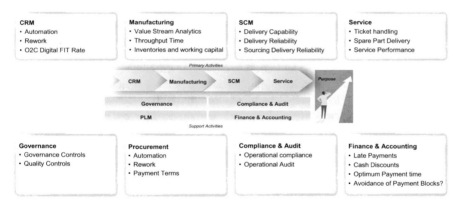

Fig. 3.2 Operational use cases

made to the previous value chain structure, as Procurement has proven a more common use case than PLM, so that PLM will be replaced in the following discussion by Procurement. Most of the following use cases will be complemented by a more detailed discussion presented by experts in Part II.

CRM

CRM is a prime area where Process Mining can add significant value. Typical driving factors are automation, reduction of manual rework activities, and support of digital transformation. Use cases range from O2C as discussed in the Siemens' case to standardization as discussed in the Uber case. O2C reporting has proven to be a good starting point and prime showcase for Process Mining. Event logs from order entry till cash collection are typically stored in ERP systems, where the Processtraces are relatively easy to collect. Quick insights can be gained with visualizing steps, e.g., order entry, order confirmation, delivery note, invoice, and cash collection, allowing to identify automation potentials and unnecessary rework.

Automation pursues the objective to substitute mundane manual activities with automation, e.g., in the backend system, batch processing, or by deploying RPA technology. Typical samples for activities, which are frequently performed manually, but have high potential to be automated, are "Create Sales Order Item," "Create Invoice," "Create Delivery," or "Remove Delivery Block." With Process Mining, all manual activities can be identified and assigned to the responsible unit, allowing operational experts to identify, assess, and assign potentials for automation. Classical automation is applied in the backend system with batch processing, the collective execution of several process activities. While this is widely applied, e.g., for processing large numbers of monthly invoices, Process Mining provides additional transparency which can identify further potentials for batch processing. Other levers for automation are workflow automation tools or RPA. In particular RPA allows to extend the usage of automation towards single users, as bots become deployable by individuals in order to substitute single manual activities.

Rework is unnecessary, repetitive work which causes waste of time and effort. Rework occurs, e.g., when a PO denominates five pieces at $5, but the PO confirmation states five pieces at $6. In consequence, this order needs to be "reworked," as someone needs to spend time and effort to clarify the difference. Typical samples for activities which show high rework frequencies are "Change Price," "Remove Delivery Block," "Change PO Profit Center," or "Remove Billing Block." Process Mining allows to discover and assign all rework activities, e.g., by operational entity, product, or customer, allowing operational experts to assess and reduce rework efforts.

The global O2C reporting at Siemens Industries, which is presented in more detail in Chap. 9, provided transparency for more than 30 million items ordered in fiscal year 2017. Processing these orders included more than 60 different process steps with more than 300 million single activities per year. Figure 3.3 shows that customer orders were processed in more than 900,000 different process variants, with smooth processing and high efficiency being paramount requirements for these global operations.

This kind of insight is a fundamental first step, but needs to be turned into operational action and value. The "O2C Digital Fit Rate" for customer orders is a prime showcase on how to turn insights into action. The FIT rate reflects the average number, how often a customer order has to be touched manually. For calculation of the KPI the number of manual touches required to process customer orders is divided by the total number of orders. A unit which requires on average 50 manual touches per 100 orders has a Fit rate of 50/100 = 0.5. A unit which requires on average 200 manual touches per 100 orders has a Fit rate of 200/100 = 2.0. This simple KPI provides a transparent and comparable performance factor per unit and enables management to measure and monitor progress. Strategic targets such as digitization

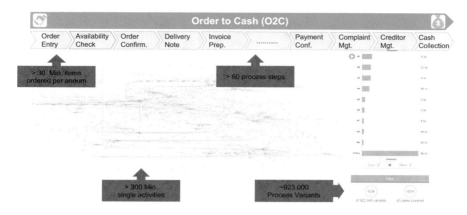

Fig. 3.3 O2C use case

of activities can be individually assigned, allowing to track progress with the process owners on a regular basis. With the same data provided from a single source of truth, operations can identify and eliminate unnecessary steps and increase automation.

In this specific example, the sector CFO started a global program to drive digital transformation of the O2C process in 2017. The Digital FIT rate was chosen as KPI for regular performance reviews with regional management, combined with targets for automation and rework rates. These global targets were linked to the local analysis, which was available to the operational order management, thus providing one single source of truth for operational order management, regional management, and corporate management (Fig. 3.4).

The project achieved impressive results with respect to increase in automation, reduction of rework, and manual activities: the overall automation rate could be increased by 24%, rework could be reduced by 11%, and a total of 10 million manual

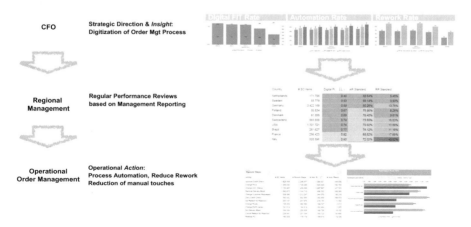

Fig. 3.4 O2C single source of truth

Fig. 3.5 O2C results

touches could be eliminated. The overall achievements were showcased in the Gartner IT Symposium 2018, as shown in Fig. 3.5.

Manufacturing

Manufacturing is a more challenging field for Process Mining. Most manufacturing processes are unique per plant, with heterogenous data sources, which makes scaling across plants difficult. While Process Mining is challenged in a fully automated production environment, semiautomated plants may well take benefits for their value stream optimization, as showcased in the BMW use case in Chap. 11, providing an example for a paint shop. In the following, we will focus on two use cases, for value stream analysis and inventory optimization.

Value stream analysis can be supported based on event logs from ERP or Manufacturing Execution Systems (MES), reflecting digital traces from actual manufacturing steps. Each single activity performed by a machine during the manufacturing process leaves a digital trace. Collecting and visualizing these traces with Process Mining allows a better understanding in the form of a digital twin of the actual value streams, including activity sequences and times, as shown in Fig. 3.6.

The classical approach for a value stream analysis in a production environment requires manual documentation of single activities, a time-consuming effort which can only represent a subset of events. Process Mining allows a complete transparency of all events, which can also be provided in near-real time for quick optimizations.

Use cases have been applied in several plants and include analytics of line jumpers, bottlenecks, and optimization of throughput times. Line jumpers may

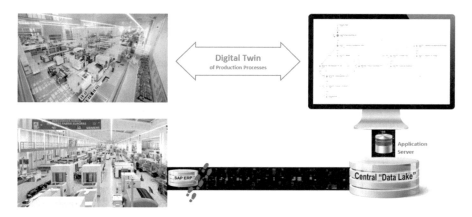

Fig. 3.6 Digital twin in manufacturing

occur with parallel production lines, when material "jumps" between lines. Reasons can be machine outages or bottlenecks in one line causing material to get stuck. While classical analytics does not provide time dimensions, which allow identification of actual flows, Process Mining can provide actual timestamps, thus showing the manufacturing sequence. The same applies for bottlenecks, which delay the value stream in a production line as material is stuck with single machines.

Throughput times typically represent a crucial factor for production efficiency. Process Mining not only allows to measure overall times, but also dependencies, e.g., regarding different materials and simulations for load optimization.

Inventory and working capital are key factors for financial efficiency. In a manufacturing environment, where fixed assets in the form of raw material and finished goods in different plants and locations can easily accumulate to a significant volume, Process Mining allows insight into actual material storage and optimization potentials. A digital inventory management takes, e.g., ordering or storage times into account for insights regarding the processing time from material ordering till delivery. It allows for in-depth analysis of inventories with materials clustered depending on turnaround times, thus allowing for optimized ordering processes, reduction in stock, optimization of working capital, and reduction of wasted material.

Logistics

Process Mining is applicable in logistics both for the sourcing transport from supplier to manufacturer as well as for the delivering transport from manufacturer to customer. Throughput times and on-time deliveries are of critical importance for customer satisfaction and cost efficiency. Performance improvement of the e2e deliver processes must be based on transparency throughout the whole logistics process and helps to identify bottlenecks, delays, and to assure on-time deliveries.

In a traditional performance reporting, the above-mentioned KPIs are typically reported by each unit individually once per month. Multiple data sources with diverse data models leave room for individual interpretation and inconsistent reports, thus making a globally consistent management difficult. Process Mining allows for a new dimension of fully automated performance reporting based on one single source of truth.

Delivery Capability is a typical KPI for the logistics performance management and measures the capability to confirm the delivery date requested by the customer with the first confirmed delivery date. With each order confirmation, a delivery date must be committed, which can depend on multiple constraints in order processing, manufacturing, distribution centers, and transport. Furthermore, the confirmed date can shift during time, with a negative impact on the delivery capability. Process Mining allows to measure this KPI and identify changes.

Delivery Reliability indicates the reliability to meet the first confirmed delivery date with the actual order fulfillment date. Process Mining allows to measure all relevant events and activities, e.g., order confirmation, delivery note, on-site delivery, confirmed delivery date, and visualizing the actual process flows. Reasons for delayed deliveries can be identified and remediated. One concise reporting allows the whole global organization to rely on validated data and focus on performance improvements.

Sourcing Delivery Capability is a critical success factor for supplier performance and supplier management. Insight into the actual performance of every single supplier allows objective discussions with suppliers, identification of low performers, and reasons for late supplier deliveries.

Especially in logistics, Process Mining has been acknowledged for supporting a more efficient and sustainable economy. Logistics experts have been able to increase supply chain efficiency by reducing transportation routes, optimization of inventories, and use insights as a perfect decision base for transport modal changes. Process Mining provides transparency and insights which allow to reduce transactional cost and environmental footprint, thus creating economic and ecologic value.

Service
Service has proven a more challenging field for Process Mining, primarily due to heterogeneous processes and a variety of source systems for tracking event logs. Though service processes—on a very high level—appear homogeneously designed across businesses, the operational handling of service events typically depends on local adoptions and requirements. Transparency can facilitate standardization of heterogeneous service processes, as the Uber use case shows in Chap. 10.

Ticket Handling is a typical use case, where Process Mining allows insights to understand slow handling of tickets and delayed customer service. Service tickets might be delayed due to clarification, approvals, or process blocks which can be disclosed with Process Mining.

Spare Part Delivery often needs to be assured under time-critical constraints with delays impacting customer satisfaction. Established use cases identify late deliveries and assure on-time deliveries of spare parts, and thus help increase the customer satisfaction.

Service Performance can be measured in different dimensions, with Process Mining providing the additional valuable information of time sequences and process flows, which most other tools are not able to provide.

Governance
Though governance is not a primary field for Process Mining projects, *governance controls* can be operationalized with the insights based on actual event logs. Global implementation of regulations can be monitored with respect to status and progress. Corporate management can be supported with corporate reporting to manage a global organization. As Process Mining allows continuous monitoring, continuous improvement programs can be supported with progress monitoring. Equally, *quality controls* can be visualized and tracked with Process Mining.

The ABB use case in Chap. 16 provides specific examples, where Process Mining is used in a governance model to drive business transformation in Divisions and Countries, supported by a Center of Excellence (CoE) and Group Quality & Operations. It describes the shift from mining processes to driving processes, accompanied by a continuous improvement program. The Bosch use case in Chap. 17 presents Process Performance Indicators, which allow top management to track process improvements. In the Schukat use case in Chap. 18, Process Mining supports the whole organization to reinvent itself with a data-driven approach.

Procurement
Procurement has seen a wide adoption of Process Mining use cases for P2P. Key drivers are easy access to relatively homogenous data, as event logs are typically stored in ERP systems. In addition, many procurement departments have a high priority to reduce transactional costs and digitize processes, so the "People" factor is favorable. Similar to O2C, P2P has a high focus on automation and reduction of rework and seen some valuable benefits resulting from Process Mining insights.

Automation of procurement activities focuses on substituting manual activities with automated processes, either directly in the backend system or by using RPA technology. Typical activities with high automation potentials are "Send PO," "Record Invoice Receipt," or "Pay Invoice." Process Mining provides transparency regarding manual steps and times required for processing, thus allowing to prioritize automation potentials.

Rework occurs when a PO requires to be reworked, e.g., due to "PO update," "Change Delivery Date," or "Change Price." These rework activities induce unnecessary workload on the organization. Rework rates can be visualized, e.g., by vendor, product, or operational unit, thus providing transparency to those in charge of process optimization.

Payment Terms has proven to be a use case which is easy to implement with immediate financial benefits. Payment term deviations result from diverging terms defined on the POs and the PO confirmations. As an example, a PO sent to the supplier might state "90 days net." The order confirmation from the supplier states "0 days net," which means that the payment would be due immediately upon delivery. Transparency regarding these deviations allows corrective action, resulting in later payments and improved working capital. Payment terms is a classical case where process assumption ("To-Be") and process reality ("As-Is") differ: assumptions would typically be based on circulars and master data, requesting standard payment terms. However, operational reality often shows significant deviations as shown in the example above.

Further use cases have been established with procurement and will be discussed in detail and with first-hand experiences in the Siemens P2P use case in Chap. 12.

Compliance and Audit

Compliance would intuitively appear as a prime domain for Process Mining: insights into the time sequences of activities allow easy detection of compliance issues such as payments without POs or segregation of duties. Detective controls can be applied based on the actual digital traces, even for activities and processes which are distributed across different source systems. However, compliance has not generally emerged as a field where Process Mining is widely applied. Expert discussions and published use cases to date seldom refer to cases where Process Mining has added value for compliance transparency.

Audit—on the very contrary—has been a focus area right from the beginning of Process Mining, and the Bayer use case presented in Part II is a prime example. Audit departments were amongst the first to use Process Mining to support projects with fact-based assessments and discussions. Data insights are used to prepare audit projects in regional companies and operational units. For a typical project, a one-time data extraction from a local source system would be processed on an SQL database and allow to understand operational efficiency across different functions, thus preparing the on-site audit project in a more profound and quantitative manner. Use cases have evolved significantly since then, as reflected in the Bayer use case in Chap. 20, where Process Mining has been used for many years by the central audit team as a technical foundation for digital insights, and lays the foundation for the digital future of internal audit. Process Mining supports the audit of process conformance, assessment of process deviations, and identification of compliance issues. e.g., with respect to segregations of duties. Many of the large audit companies have started using Process Mining to improve the reliability of their audits.

Finance and Controlling

In finance and controlling, accounts payable (A/P) as well as accounts receivable (A/R) have seen significant improvements with Process Mining. Transparency supports on-time payments, securing discounts, and overall process efficiency.

Late payment of A/P can cause loss of cash discounts or result in penalties. Late payments can easily be flagged out by combining the event logs from POs, PO confirmations and payment activities. Predictive alerting allows a proactive information about upcoming or overdue payments. And in some particular cases, regional companies, such as in South Korea, might even be legally required to show proof of punctual payments.

Cash Discounts incentivize on-time payments and can provide financial benefits. Discounting payments requires transparency about the timestamps and status of each PO and PO confirmation, which can be visualized with Process Mining and even be proactively alerted to users.

The *Optimum Payment time* may depend on a variety of different factors, e.g., payment terms, discounts, liquidity, month end, and payment transfer times. As Process Mining allows to visualize the time sequence of each payment, it can provide valuable insights supporting optimum payments.

The Telekom use case in Chap. 21 provides experiences on how Process Mining is used in a multi-shared service center in charge of procurement, accounting, HR, and reporting services. The Procure to Pay process is discussed in detail.

Further Use Cases
The list of use cases discussed above is not exhaustive, as there is a large number of additional functions where Process Mining has been applied, and only a small selection shall be mentioned here:

Process Mining allows a better understanding of the Product Lifecycle Management (PLM) process. It helps to reduce development cycles and identify potentials to establish or improve standards. Besides thorough understanding of time sequences between single activities, a predictive alerting of events such as service renewal date can been visualized. Siemens Healthineers presents a use case in Chap. 19, where Process Mining allows to understand how customers work with computed tomography devices, thus providing product management with valuable insights.

Human Resource Management can be supported, e.g., for insights regarding the hiring process. Smooth processing of applications, interviews, and onboarding are examples of operational use cases where Process Mining not only provides transparency on throughput times, but can also provide insights into selected cluster, e.g., male/female, age, etc. Insights across the whole Hire2Retire cycle could also be of interest, but usually face the challenges of insufficient digital processtraces and restricted data access.

Banking services are using Process Mining to track turnaround times for credit approvals, from credit application till approval and receipt of funds. Tracking the e2e process not only allows to monitor the status of each application, but also to identify reasons for delay. Insurances use Process Mining to track the underwriting process in a similar form.

ServiceNow is an example for an application which can be screened for delayed issues. Service tickets managed in ServiceNow can be traced regarding their handling timeline and Process Mining supports to identify delays, understand reasons for bottlenecks and speed up overall throughput times.

Experience has shown that there is an unlimited field of potential use cases. Based on individual domain know-how, people should be enabled to apply this digital tool in their particular environment, which makes People a crucial aspect.

People: The Human Factor

Lars Reinkemeyer

Abstract

While innovative IT tools can be a great enabler, it is a key success factor to get the right people on board. All smart data, insights, and transparency will be useless if the process experts or process owners do not appreciate and support the approach. Similar to applying MRT technology, the affected people must be determined to pursue a therapy and strive for improvement. The chapter shares operational experiences, challenges, and organizational setups which have proven successful.

"Technology is nothing. What's important is that you have a faith in people, that they're basically good and smart, and if you give them tools, they'll do wonderful things with them." Steve Jobs.

While innovative IT tools can be a great enabler, it is a key success factor to get the right people on board. All smart data, insights, and transparency will be useless if the process experts or process owners do not appreciate and support the approach. Similar to applying MRT technology, the affected people must be determined to pursue a therapy and strive for improvement. The process owner must be seeking digital transformation or process improvements and actively support the usage of Process Mining.

> *Key Learning #6: It's All About the People*

Having the right people joining on this exciting new journey can reveal the magic of human collaboration. Any organization which is challenged in a changing

L. Reinkemeyer (✉)
University of California, Santa Barbara, Santa Barbara, CA, USA
e-mail: reinkemeyer@ucsb.edu

© Springer Nature Switzerland AG 2020
L. Reinkemeyer (ed.), *Process Mining in Action*,
https://doi.org/10.1007/978-3-030-40172-6_4

environment, e.g., digital transformation, requires creativity and collaboration to use new digital tools. An iterative approach allows to jointly learn what can be made transparent and how insights can yield value. The right approach can provide an environment that unleashes human potential as a differentiating factor.

In an ideal environment, the journey starts with the right process owner who is passionate about change and innovation, actively supporting this new technology and helping to focus on the most urgent operational challenges. As insights and value become more transparent, the number of people supporting the approach will continuously increase. Look for people who are open to a season change instead of conserving the status quo. Process Mining can provide insights which allow to question established assumptions, procedures, and processes. Change can be achieved with people who know that there are things they don't know. People who are open to things they don't know that they don't know—the unknown unknowns. Successful projects rely on curious insiders with the right mindset and in-depth understanding of the respective processes, organization, and people. Search for these kinds of change agents and make them process heroes by supporting them with the right tools and by acknowledging them for the achieved impact. The Siemens O2C use case in Chap. 9 provides a perfect example of a lean project team, with the right skills and passion. EDP discusses skill requirements in Chap. 15.

At the same time, be prepared to encounter experts who have managed a process for many years and who are reluctant to change. Process Mining can reveal inefficiencies which have been previously neglected or ignored. Classical work structures and process design will be challenged and people might be obliged to justify the way they have worked in the past. They might even be scared of their future contribution or of being substituted by automation. When employees perceive Process Mining as a potential threat to their jobs, they may consciously or unconsciously resist transformational efforts. As transparency can be perceived as a threat, this should be proactively addressed, e.g., with an open communication of targets, early involvement, and continuous training.

As described in Chap. 3, a purpose needs to be defined. If people understand the overall purpose and how insights and metrics generated with Process Mining support to achieve this purpose, the full impact can be yielded. If there is misalignment between purpose and approach, surrogation may occur: people start to question the insights, focus on tricking the metrics instead of supporting the purpose. If, e.g., the confirmed delivery date can be easily adjusted in the source system to improve the KPI Deliver Capability, focus will be on date adjustments to improve KPIs, rather than supporting the purpose of on-time deliveries. People at all levels have to trust the analytical results and suggestions, with a clear understanding of how and why working with the data helps to achieve the overall target and purpose.

While technological change is often perceived as a threat, it can also be an opportunity for new skills and roles. Many of the roles described above have led to new types of experts, both in business and IT. Harvard Business Review has, e.g., coined the role of "analytics translators,"[1] who bridge the gap from technical realms of data engineers and scientists to the people from the business realm—procurement, supply chain, manufacturing, sales, finance, etc. This kind of competency is of increasing value and offers new opportunities. As a general trend, the automation of work is continuously progressing and part of the economic evolution. At the same time, it goes against human nature to work on monotonous tasks, and robots take over repetitive or mundane work, while leaving opportunities for new tasks and skilled workers.

Regarding the organizational setup, a close collaboration between functional departments and IT is recommended. The staffing of a project team should include people with deep process know-how as well as team members with IT competencies, who are required to identify and customize the digital traces. The Siemens O2C use case in Chap. 9 shares first-hand operational experiences from a global rollout project. Once Process Mining has been commonly accepted as a new technology in an organization, it should be embedded in the organizational structure. Figure 4.1 shows a setup which fosters an intense collaboration between functional departments and IT. Digital transformation, operational excellence, and Process Mining are tasks which are typically on the agenda of the process owners and requires a strong technical support. To bridge the gap between business and IT, dedicated people should be nominated on both sides. This can be achieved by the nomination of people or teams for each application, e.g., in procurement finance, logistics, or sales. The nominated "twins" from business and IT should work closely together towards joint targets and develop a good understanding of the counterpart's environment.

On the IT side, the application owners can become entrepreneurs, responsible for the global application development, stable operations, technical architecture, innovation, rollouts, financial planning and budget, usage, and strategy. Agile collaboration has proven efficient for these teams to intensify collaboration between

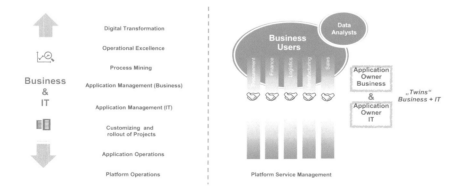

Fig. 4.1 Organizational setup

[1]"How to Train Someone to Translate Business Problems into Analytics Questions," HBR 02/2019.

business and IT. At the same time, the agile approach has shown good results in collaboration within IT teams, which are regionally distributed onsite, nearshore, and offshore. Customer proximity requires the application owner IT to be close to his business twin and application users. Customization and rollout would typically be managed from nearshore and supported by an operations team which is located offshore.

Several use cases in Part II present dedicated CoEs for Process Mining, e.g., BMW, where the CoE assures—amongst others—process excellence, tool experience, and operational excellence. The Telekom use case in Chap. 21 discusses the roles and responsibility of the CoE in detail. Athenahealth (Chap. 14) has established a Process Mining Insights Team with selected experts driving the topic throughout the organization. At Bosch (Chap. 17), the application ownership has been assigned to in-house consulting, with strong top management support. Complementary to these samples with strong central ownerships, organizational setups can be built on a hub and spoke system, with the hub represented by a central competence team and spokes in the operational units.

The application teams should be supported by a platform service management, which is in charge of the application, stable operations, technical architecture, authorization management, change request handling, financial planning and budgeting, platform roadmap, usage, and strategy. An active community might be led by a service management board, where application owners from business and IT regularly discuss points of strategic and operational relevance, supported by senior management. A wider Process Mining community should be established with regular information updates across various social media channels, training for data analysts and community events.

Processtraces: Technology

5

Lars Reinkemeyer

Abstract

Processtraces are comparable to raw oil: they are hard to find, the collection causes technical challenges, and the refinement is laborious. But once all these obstacles have been overcome, it can be used in an amazing variety of forms and fuel impressive results. The chapter discusses not only best practices for identification and customization of processtraces from raw data, but also a structured approach, which includes technical aspects and challenges. The chapter covers technical architecture as well as user-related aspects such as data access, security, and user experiences.

Processtraces are comparable to raw oil: they are hard to find, the collection causes technical challenges, and the refinement is laborious. But once all these obstacles have been overcome, it can be used in an amazing variety of forms and fuel impressive results.

As a first step, the identification of relevant digital traces and customization of reliable data models is required, which typically accounts for the majority of work. In the following, we will discuss not only best practices for identification and customization of data, but also a structured approach which includes technical aspects and challenges.

Processtraces are the technical foundation of Process Mining. Insights are built with digital traces which have been extracted from backend source systems. These

L. Reinkemeyer (✉)
University of California, Santa Barbara, Santa Barbara, CA, USA
e-mail: reinkemeyer@ucsb.edu

© Springer Nature Switzerland AG 2020
L. Reinkemeyer (ed.), *Process Mining in Action*,
https://doi.org/10.1007/978-3-030-40172-6_5

traces are customized and provided to the users in an intuitive frontend, allowing the users to identify exceptions and complexity drivers. The approach has become possible with the increasing power of processing "Big Data," large amounts of digital traces and powerful frontend tools which allow fast presentation of analytical data. Operational reports may include millions of events and billions of data elements, posing a continuous challenge for fast processing and visualization, but—when accomplished—fuel amazing insights.

> ### Key Learning #7: Processstraces Are Comparable to Raw Oil

The technical foundation for a Process Mining project can be described in a multistep approach: the first step is data ingestion, which includes the discovery and extraction of relevant processtraces and facilitates data flow from any source system into one storage environment. While the relevant event logs are determined by the purpose—i.e., which activities are to be considered—the identification of the event logs in the source systems can be a major effort driver. It not only requires a clear understanding of which log is stored where in the data base, but also access to the all necessary data. In large, global organizations digital traces may be distributed across multiple, different IT sources including ERP and non-ERP systems. Typically, the individual customization of source systems is a major effort driver as also described in the ABB use case in Chap. 16. Consistent master data management can facilitate homogenous raw data, resulting in less effort for customization. However, experience shows that in most companies there is only little consistency in those master data which contain the event logs required for Process Mining. Several vendors have started using AI and ML to compensate these deficiencies, with the promise of automatically identifying and normalizing event logs from various sources. As a challenge, bias and noise in raw data must be properly compensated and most solutions have not yet proven sufficient in the face of experienced process experts looking in detail at the data.

A dedicated methodology can support the data collection, with templates in the form of wanted posters, which are used to discover and document relevant event logs, system data, and named contact partners. Once the event logs have been identified, the technical infrastructure should allow a direct ingestion of all relevant event logs. This Extract, Transport, Load (ETL) approach can be supported with SAP Landscape Transformation (SLT) or Smart Data Integration (SDI) replication technology.

Similar types of event logs are typically stored in the source systems in various tables or fields, due to individual system customization. They may be stored in different sources, such as ERP systems (e.g., SAP, Oracle) or workflow management systems (e.g., Pegasystems, Salesforce, ServiceNow, Windows workflow foundation). In order to show apples for apples, it is required to standardize and harmonize the ingested raw data. This allows to visualize digital traces, which have been ingested from different source systems, as a seamless e2e process, showing the user the complete process flow.

Discovering and standardizing the event logs is a major effort driver and it has proven valuable to reuse this customized "Smart Data" in further projects. As single event logs may be required for the visualization of logistical, financial, and sales analytics, they should be provided in an open architecture for other projects to avoid double effort. The "refined" data can be provided to multiple projects with a semantic layer. As an additional effect, this provision of standardized event logs in a joint semantic layer assures one single source of truth and consistent data analytics across use cases and functions, both being of essence for user acceptance.

As a final step, the normalized event logs are connected to actual process flows and visualized in frontend reports, which allow insights for business users. A powerful frontend tool must be able to show single activities, process flows, and all process variants in an intuitive manner and with fast updates for drill downs. The market offers a variety of different Process Mining applications for visualization and analytics with a range of differentiators. Typical decision criteria for these applications are intuitiveness and usability, performance and stability, scalability and innovation, references and price.

The different level of a typical technical architecture are shown in Fig. 5.1:

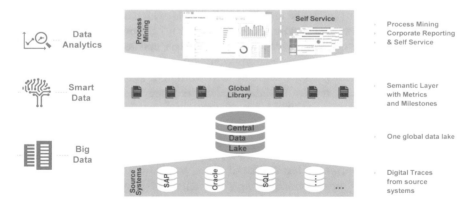

Fig. 5.1 Architecture

With a growing demand for data democratization, self-service fuels individual process analytics and is of increasing importance for user acceptance. Once a semantic layer has been established with standardized event logs, users can be granted access to build individual analytics. Users can either develop individual reports with predefined data cubes or be enabled with a few days of training to conduct independent data analytics using the standardized logs provided in the semantic layer. For an open knowledge platform, trainings and community meetings allow to exchange experiences and best practices.

Key Learning #8: Provide an Open Platform to Build a Strong Community

Independent of all technical and architectural challenges described above, the user will expect 100% stability and immediate availability of reports. Before a user is provided with analytics via smart interfaces, complex procedures are executed, with billions of activities and complex calculations in the background. The calculation of millions of event logs from several source systems in one central database can take up to several hours, as billions of data records need to be processed. In a classical on-premise environment, the calculated results need to be transferred to the application server, which updates the reports for visualization in the web frontend. To track the actual data load status, dedicated reports should be provided, thus informing about data actuality and availability. To assure data consistency, the application teams should check data completeness in a regular and automated manner. Despite this technical complexity in the background, the typical user will always request reliable and fast frontend visualization providing insights in less than a few seconds.

How often a report should be updated is a hot topic, which is regularly discussed. Should it be real time, near real time, daily, or monthly? Experience shows, that the vast majority of reports is provided with a daily data update, which matches the expectations of most users. Business users would typically expect their insights to be available in the morning, e.g., when they start working a morning shift in a plant. Only a few use cases require a higher update frequency or even near-real-time reporting, which might be required, e.g., for steering a production line on a shop floor.

Security and authorization requirements are crucial aspects of any architecture. As a general trend, availability of data to a wide range of data analysts has become increasingly accepted in the last couple of years, as data has become more easily accessible and the organizational hunger for data has grown. Data protection requirements typically would differ depending on the type of data, e.g., with financial data being more sensitive while logistical event logs might be provided to a broader user community. Access to data should generally be handled on a need-to-see basis, which can be organized via authorization concepts. In the reference model described above, this can be implemented on different levels: the "Big Data" lake, which contains unfiltered raw data, would typically not be accessible to any user. Developers can request access to subsets of data on a need-to-see basis, which must clearly define the purpose of the project. Approval to the schema should be granted by the responsible data owner or authorizing deputies. The developer may use the data, which is provided in a dedicated project schema, to build a specific user report. Access to reports can be granted per user and should be continuously reevaluated.

While the first Process Mining projects were built on premise, cloud-based data collection and processing has been a major trend in the last couple of years and permeates continuously, as it offers a range of advantages: vendors can easily update their software, which is necessary at a regular frequency and imposes planning and effort for on-premise installations. Vendors can continuously monitor processing performance and adjust whenever required. And use cases like Uber show impressive turnaround times for cloud-based applications. As a downside, many customers are still reluctant to store data outside the corporate firewall in a cloud offered by a third party.

Challenges, Pitfalls, and Failures

6

Lars Reinkemeyer

Abstract

As the evolution of Process Mining has not always been on the happy path, the idea of learning from failure has been adopted as a guiding principle for this book, applicable to this chapter as well as to the use cases in Part II. This chapter presents ten samples, reflecting challenges which were posed, pitfalls which were learned hands on, and failures which have been experienced. Samples range from data availability to process conformance checking and shall help the reader to avoid similar experiences.

"Where did you fail?" was one of the first questions which I was asked during a faculty meeting at the University of California, Santa Barbara (UCSB). "What an awkward question," was my immediate thought, but then I understood that the PhD candidate wanted to learn from my experiences and failures. Unlike the typical European approach, the American approach sees failure as the mother of success. This idea of learning from failure has been adopted as a guiding principle for this book, applicable to this chapter as well as to all individual use cases in Part II, where every contributor had been invited not only to share success stories, but also to share lessons learned, including experiences on how to fail fast or scale fast.

> *Key Learning #9: Fail Fast or Scale Fast*

The selection of the following ten samples is not exhaustive, but reflects challenges which were posed, pitfalls which were learned hands-on, and failures which have been experienced. Some of the samples bear great ideas, which would

L. Reinkemeyer (✉)
University of California, Santa Barbara, Santa Barbara, CA, USA
e-mail: reinkemeyer@ucsb.edu

© Springer Nature Switzerland AG 2020   37
L. Reinkemeyer (ed.), *Process Mining in Action*,
https://doi.org/10.1007/978-3-030-40172-6_6

not fly in the particular environment, but which might create value if applied in the right environment and with the right combination of the 3Ps.

- *Data availability*: This is probably the biggest challenge for every project and has several aspects. As a first point, access to event logs poses a technical and a data protection challenge. Typically, data is hidden in the source system(s) and contains confidential information such as supplier or customer information, which the host might be reluctant to provide. The workers council might need to be involved, potentially imposing further challenges and restrictions to data availability. The second aspect is the completeness of data, which is required to visualize any business process in its totality. If digital traces are not or only partially stored in the source system, a concise reporting will not be possible. The third aspect is the continuous availability of event logs with minimum latency, which is required for regular and on-time updates.
- *Benefit calculator*: Based on the transparency and insights achieved with Process Mining, it is quite easy to assign time savings and values for certain activities. A typical assumption to calculate a business benefit could be that the automation of an activity "Send PO" can save x minutes worth y euros. Parameters and values are easily available for process experts and can be provided with calculation interfaces which allow to simulate how much benefits can be achieved, e.g., by avoiding rework, reducing approval steps, or securing discounts. While this provides an easy calculation of saving and potentials, its adoption in operational departments across the industries has been slow. Commercials and controllers tend to appreciate this approach, but process owners turn out to be rather reluctant with respect to the calculation of saving potentials in their respective area of responsibility.
- *RoI calculation*: The question of how to calculate digital benefits and the Return on Investment (RoI) of a Process Mining project comes up regularly. There are easy ways for calculation, if there is a willingness to make some assumptions and assign specific values for calculation. Positioning Process Mining as cost saving measure can turn into a disadvantage, as this cost-driven approach might bring a different focus and perception for this new technology.
- *Complexity monitor*: As complexity is a key driver for Process Mining, it becomes tempting to visualize overall operational complexity. The approach is based on the Pareto principle and calculates how many different process variants are required to process 80% of all activities. Results show that only a fraction of the total number of process variants was required for processing the large majority of events, while the remaining few activities required a huge number of process variants. The hypothesis can be applied that this large number of variants generates high complexity and leads to higher efforts. In the report, which is shown in Fig. 6.1, entities with a high degree of complexity can be flagged out and compared to units with less complexity.

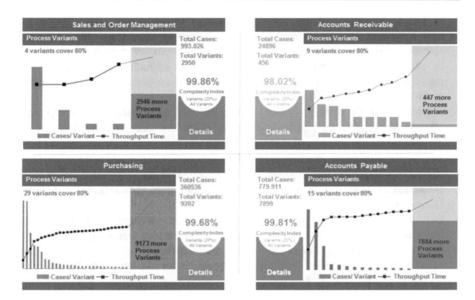

Fig. 6.1 Complexity monitor

- *Compliance processes* appear—on first glance it—to be an exciting opportunity to track malicious behavior, fraud, and compliance incidents, e.g., payments without POs. However, experience from many years and numerous discussions showed only little demand, limited value, and very few operational cases.
- *Process conformance* is one of the hot use cases, discussed intensely amongst Process Mining academics. The concept to compare To-Be processes with As-Is process flows appears compelling, as it allows a fact-based comparison of general assumptions and operational reality. Various reports have been built across the industry, e.g., for procurement, logistics, and sales, to allow experts an in-depth analysis regarding deviations between theory and reality. However, experience shows little adoption of this approach and none of the use cases presented in Part II reflects process conformance.
- For *HR* processes, a variety of use cases appear valuable, e.g., measuring a hiring on- or offboarding process (Hire-to-retire). Process Mining allows performance measuring, e.g., in call centers to visualize how many tickets are handled by single employees per day. This may be of interest in environments where individual performance measurement is commonly accepted. However, confidentiality of person-related data and heterogenous data sources pose a particular challenge for these use cases and few operational use cases are known. Responsible data usage should always be a key paradigm, and performance measurement of human activities should be critically reviewed and controlled in order to assure nondiscriminatory data analytics.
- *Project processes*: With a large number of projects, structured with standard project milestones, it appears valuable to visualize the sequences of milestones

as well as time spent between two milestones. This could allow a comparison of how long a project takes from approval till start or from finish till invoicing. However, the reports provided for project experts could not yield sufficient value, as every single incident and project delay would be explained by the project managers. In essence, no sufficient purpose could be defined and—after initial excitement—this turned into a pitfall.

- *Content store*: Quite early in our journey we tried to industrialize use cases by presenting them as standard templates in a content store accessible to every employee. While this might add value in an environment with strong focus on standardization, most departments would rather go for individual adoptions of use cases. Nevertheless, the content store was valuable as a standard menu, and presented interested colleagues a wide range of potential use cases and thus raised interest and stirred discussions.
- *Data hunger* can quickly turn into a pitfall, as process experts tend to demand more and more data. Traditional tools such as business warehouse are able to ingest and process historical data from past decades. Process Mining requires more complex calculations and visualizations, which can lead to a slowdown in speed due to increasing data volume. Typically, you will experience a constant battle between increasing data hunger and performance required to maintain high user experience.

Process Mining, RPA, BPM, and DTO

7

Lars Reinkemeyer

Abstract

With digital transformation being one of the hottest topics of today's business, digital tools such as Process Mining and Robotics Process Automation (RPA) see a spike while Business Process Management (BPM) might be considered as a more established approach for operational efficiency. This chapter describes the differences between these technologies, how they correlate and can complement each other, e.g., with the RPA Scout, and touches on the concept of a Digital Twin of an Organization (DTO).

With digital transformation as one of the hottest topics of today's business, digital tools such as Process Mining and Robotics Process Automation (RPA) see a spike while Business Process Management (BPM) might be considered as a more established approach for operational efficiency. In the following we will describe the differences between these technologies and how they correlate and touch on the concept of a Digital Twin of an Organization (DTO).

Process Mining is capable of providing a wholistic picture of all events and processes in an organization. Similar to an MRT scan of a human body, it allows perfect transparency and insights regarding actual operational efficiency. Based on these insights, weak points can be identified and decisions taken, e.g., to streamline processes, optimize backend operations, train people to perform activities in a more efficient manner, or substitute mundane tasks with workflow and task automation.

RPA complements the traditional possibilities for automation—which were primarily focused on process automation in backend systems—as it allows for operational improvement by automating single tasks or parts of workflows. RPA has enabled companies like Siemens to automate hundreds of thousands of working

L. Reinkemeyer (✉)
University of California, Santa Barbara, Santa Barbara, CA, USA
e-mail: reinkemeyer@ucsb.edu

© Springer Nature Switzerland AG 2020
L. Reinkemeyer (ed.), *Process Mining in Action*,
https://doi.org/10.1007/978-3-030-40172-6_7

hours and shifts the boundaries of process automation in respect to investment and usability. Traditional automation was backend centric, highly technical, cost intensive, and for experts only. Modern RPA technology can be applied individually by single employees to automate mundane tasks, and it enables a virtual workforce automation. Vendors provide a range of different solutions, from workflow automation stacks to simple-to-use bots. RPA automation appears highly flexible and can deliver short-term successes by deploying bots, which automate single tasks or simple work sequences. While RPA thus allows to yield quick results, it can't provide transparency regarding the overall processes, interactions, and implications, which would be necessary for a wholistic transformation.

BPM is a disciplined approach to identify, design, execute, document, measure, monitor, and control both automated and nonautomated business processes to achieve consistent, targeted results aligned with an organization's strategic goals.[1] BPM has traditionally been applied for process modeling and has proven valuable for many organizations for designing, structuring, and documenting of business processes. BPM and Process Mining differ fundamentally: while BPM allows to model To-Be processes in order to describe how processes should work, Process Mining visualizes As-Is processes and reality (Fig. 7.1).

The following analogy stresses the difference: going on a journey requires appropriate planning. Events and activities should be thoroughly structured, in order to plan the best possible sequences and flows. BPM is the perfect tool for such a process modelling and planning of the "To-Be" process. Once the journey starts, deviations will be encountered, as particular incidents might require flexible reactions. Operational reality often leads to variable process flows and unexpected deviations, which are of particular interest, as they might turn into complexity drivers. Process Mining is the perfect tool to visualize these "As-Is" process flows. The synergy of both approaches appears compelling: designing processes with BPM and checking with Process Mining the actual process conformance. Several vendors in the market offer products which are based on this value proposition.

Fig. 7.1 Plan versus reality

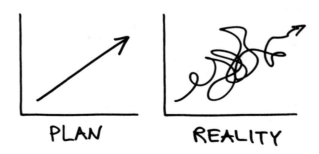

PLAN   REALITY

---

[1] https://www.abpmp.org/page/BPM_Profession

Process Mining and RPA both support digital transformation and—from a complementary perspective—can amplify each other. Bill Gates, founder of Microsoft, once said: "The first rule of any technology used in a business is that automation applied to an efficient operation will magnify the efficiency. The second is that automation applied to an inefficient operation will magnify the inefficiency." The challenge for RPA automation is the identification of automation potentials in efficient operation and to avoid automation of inefficient operations. To overcome this challenge, the event logs and process insights available with Process Mining can provide structural and operational value for RPA. Structural value is generated with Process Mining by providing wholistic transparency for reviewing and optimizing Process Models. Operational value can be generated with the RPA Scout, which has been designed to use data prepared with Process Mining as a guide to identify sweet spots, where using RPA has proven applicable and might create further value.

> *Key Learning #10: Process Mining and RPA Can Complement Each Other*

Based on the full transparency of single activities it not only allows to identify which activities are already automated with bots, but also to identify further potentials for automation with bots. Figure 7.2 gives an example for this approach: as a first step all activities currently performed by bots are flagged out (1). This is technically supported, as every activity should be logged with an unique identifier for compliance reasons, which allows to select activities performed by a bot. In a second step, the identification of units (e.g., countries, business units, plants) currently using bots allows to identify experienced units, which have developed best practices (2). The number of manual activities per unit indicates how many of those activities, which have already been automated elsewhere, are conducted manually (3). In the last quadrant the RPA Scouts allows to identify the potential for bot

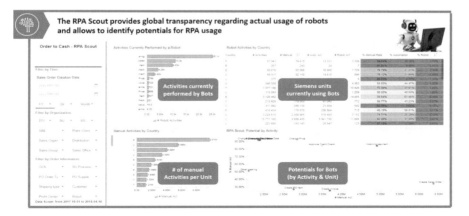

Fig. 7.2 RPA Scout

automation for each single activity, with the dimensions of total number of activity and current degree of automation (4).

Remarkably, the perception of Process Mining and RPA strongly differs in Europe and USA: a market study conducted by IDG Research[2] in 2019 with more than 350 European companies shows an equal evaluation of Process Mining and RPA with respect to impact, value, and planned investment volume. The perception in the US market differs dramatically, with RPA hyped and Process Mining rather recently emerging as a hot topic. One possible explanation might be that Europe has a long history of BPM and Process Mining. Academic research for Process Mining in Europe has established a large community and many young entrepreneurs who are hungry to convert theory into praxis, either in established organizations or by starting exciting new companies. In the USA, this foundation appears to be missing and the American management approach appears to be in favor of quick results, which are better supported with RPA, while the wholistic approach of Process Mining requires a more extensive approach.

Looking at the bigger picture of Process Mining and RPA reveals that both markets appear still very small according to Gartner's estimates: for 2018 the market volume for RPA was estimated at $850 Mio and for Process Mining at $160 Mio.[3] Remarkably the two categories have brought up four unicorns, with each of them valued much higher than the combined market volume. Something must be exciting about these technologies!

The concept of Digital Twins of an Organization (DTO) complements the established concept of Digital Twins, which are commonly used for products, plants, airplanes, cities, and many more. According to Gartner, a DTO is "a dynamic software model of any organization, that relies on operational and/or other data to understand how an organization operationalizes its business model, connects with its current state, responds to changes, deploys resources and delivers expected customer value."[4] A DTO is a virtual copy of an actual process, which facilitates analysis and understanding. The comprehensive twin of all process activities and complexities takes expert discussions to a new, unbiased level and supports the strategic management of an organization with fact-based insights. Strategic directions such as digital transformation programs become measurable and thus easier to manage.

---

[2]https://www.computerwoche.de/a/process-mining-und-automatisierung-sind-schluesselelemente-der-digitalisierung,3547348
[3]Gartner "Market Guide for Process Mining," 06/2019, Page 13.
[4]Gartner "Market Guide DTO 2018."

Key Learnings

8

Lars Reinkemeyer

Abstract

As a summary for Part I, this chapter comprises the ten key learnings on one page.

#1 Transparency is a prerequisite for digital transformation: Digital transformation should be unbiased, transparent, and measurable. Process Mining allows transparency on actual processes and provides insights for fact-based digital transformation.

#2 Process Mining allows full transparency based on event logs: An event log is a collection of events which have taken place in order to perform a business process. Event logs allow to visualize process variants in the actual sequence of activities.

#3 Purpose, People, and Processtraces are essential ("3Ps"): All three factors have to be in sync for successful projects and can't substitute each other.

#4 Start with simplicity to fight complexity: Clear and simple project targets should be defined to allow for a successful start. After a successful start, the scope can be extended to achieve transparency on process complexity.

#5 Purpose comes first: at the beginning of any project a purpose with a clear business value must be defined, typically in the form of a use case. The purpose must be achievable and supported by the process owner and all involved team members.

#6 It's all about the people: Engaging the right passionate drivers in operational units, who understand how Process Mining can improve operational efficiency and drive digitalization, is crucial.

#7 Processtraces are comparable to raw oil: you have to search for it, find it, collect it, and refine it. In the right quality, it can become a very powerful fuel.

L. Reinkemeyer (✉)
University of California, Santa Barbara, Santa Barbara, CA, USA
e-mail: reinkemeyer@ucsb.edu

© Springer Nature Switzerland AG 2020
L. Reinkemeyer (ed.), *Process Mining in Action*,
https://doi.org/10.1007/978-3-030-40172-6_8

#8 Provide an open platform and build a strong community: Process Mining services should be accessible to every employee, with an open exchange of experiences.

#9 Fail fast or scale fast: An agile project setup should allow quick results and checkpoints. Fast scaling of services should be considered in every project phase.

#10 Process Mining and RPA can complement each other: As both approaches pursue the same target—to support digital transformation—complementary approaches are recommended, e.g., the RPA Scout.

Part II

Best Practice Use Cases

To provide a wide range of experiences and different use cases, the editor has invited contributors from multiple industries (e.g., Manufacturing, Telco, Healthcare), functions (e.g., CRM, SCM, Shared Services), and different company sizes. The book is vendor independent and the following use cases have been built with Process Mining software from different vendors.

The following 12 use cases reflect the broad scope, how Process Mining can create value for different functions along the value chain):

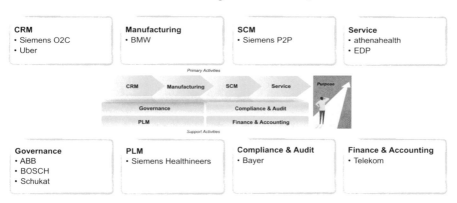

Fig. II.1 Use cases

To assure a concise reading, all use cases are structured with the following points:

- Abstract: Brief summary of the case.
- Challenge: What was the initial problem?
- Use Case: Description of the use case and how it was implemented.
- Impact: Which (measurable) results have been achieved?
- Technology: How was technology used, i.e., for backend and frontend?
- Lessons Learned: What went well and what failed? What can others learn?
- Outlook: Perspective and expectation on the future of Process Mining?
- Links: URLs, Videos, Articles, Pictures, etc.

Siemens: Driving Global Change with the Digital Fit Rate in Order2Cash

9

Gia-Thi Nguyen

Abstract

Global change that sticks in a complex organization is not an easy task, yet this has been achieved in only 1 year with a lean team of three people within Siemens Digital Industries. Using the innovative technology of Process Mining and equipped with frontline experience as well as a distinct mindset, automation and digitalization have leaped forward tremendously on a global scale. This is the call for action, because everyone can achieve the same, as the secret sauce is simply the combination of head, heart, and hands.

Challenge

Process improvement is not a new topic for any regional company of Siemens. Many organizations have undergone process mapping workshops, hired internal or external consultants and auditors, or have conducted operational reviews on a cross-functional level altogether. The success of these activities was often questionable: it was a one-time activity on top of the daily work causing heightened levels of effort which in addition were neither scalable nor truly shareable to other organizations. Furthermore, there was no easy way to make the current process landscape transparent nor show the process performance of each unit, as by the time that mapping is done, reality moved on in the form of reorganizations, change of key experts, loss of management attention, or decreasing budgets.

While regional organizations worked tirelessly on increasing automation and reducing rework activities as well as they could, it was very hard to show or prove the positive impact of these actions in a complex and ever-changing environment. Some regional companies created standardized reports, yet they were only local

G.-T. Nguyen (✉)
Siemens AG, Digital Industries, Erlangen, Germany
e-mail: Gia-thi.nguyen@siemens.com

© Springer Nature Switzerland AG 2020
L. Reinkemeyer (ed.), *Process Mining in Action*,
https://doi.org/10.1007/978-3-030-40172-6_9

viewpoints which could not be transferred to other regions in order to compare process efficiency across countries. Some global initiatives were conducted in the past from time to time, but no breakthrough improvements were able to be shown mostly due to the lack of speed or visible positive outcomes.

With a growing frustration of the regional companies in an increasingly competitive environment, it was very challenging to understand complexity while the expectation of reducing costs continuously was very real. To add to the challenge were historically grown systems that were not standardized and took a life of their own. Aligning global guidelines or targets defined by the central headquarters that lost touch with the regions resulted in contradicting messages: Is a manual activity bad if the overall cycle time is comparably faster than other countries with lower automation rates? How do we define a clean order beyond the point of order entry?

Use Case

In early 2016 top management was again eager to address the topic of O2C processes in the regions. This time there were talks about centralization of organizations, consolidation of ERP systems, or creation of best practice communities. The global program that was initiated was called Order Management for Tomorrow (OM4T) and the goals were to further automate sales back office processes in the regions and to reduce manual rework, in order to lower cost of the regional order management of Siemens Digital Factory and Process Drives Industries, now Siemens Digital Industries.

O2C refers to the processes in the regional companies from order entry, order processing, follow up, billing, and cash collection in the customer facing realm. Since all these activities are performed in ERP systems, the data trail from each system could be transferred into a global data model across all regions of all organizational units of Siemens. As a matter of fact, the technology of Process Mining has incidentally met the global program and the existing out-of-the-box data model of O2C looked very promising. After a revision of this data model with some small changes, we were amazed by the major impact it would create. As new activities were loaded into the data model while existing ones were simplified, the small and lean project team worked very closely with the internal IT organization. It was very important to always remember that our aspiration was that this data model was valuable to all regions, so we did not apply a regionalized customization but kept it global. After 2 months our final O2C data model was implemented and we were able to ramp up over 600 users in the first 6 months across over 50 countries because there were no regional rollouts necessary—O2C Monitor was available globally from the first day. Process Mining showed a new level of transparency. Indeed, it was a transparency beyond imagination. Figure 9.1 exemplifies the magnitude of our data scope—over 70 million sales order items across data from 90 countries with over 1.5 million process variants can be analyzed in the O2C Monitor that has only one standardized data model.

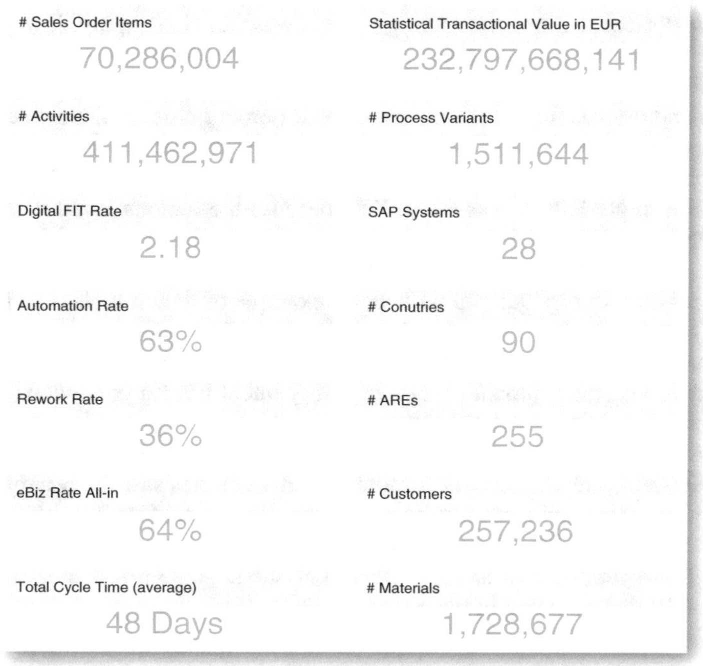

# Sales Order Items	Statistical Transactional Value in EUR
70,286,004	232,797,668,141
# Activities	# Process Variants
411,462,971	1,511,644
Digital FIT Rate	SAP Systems
2.18	28
Automation Rate	# Conutries
63%	90
Rework Rate	# AREs
36%	255
eBiz Rate All-in	# Customers
64%	257,236
Total Cycle Time (average)	# Materials
48 Days	1,728,677

Fig. 9.1 O2C key figures

Nonetheless, transparency can wow your organization, yet it will not make any process better directly. Thus, a new approach was needed to make sure that every single potential user had some benefit besides mere transparency.

The approach that was chosen was simple: in order to decrease the number of human interventions or human touches, we needed to focus on the human touch in terms of traits that define our humanity: creativity, collaboration, empathy, compassion, or imagination to just name a few. A very lean project team of three therefore embarked upon a journey to visit all lead branches of Siemens which were at that time roughly in 30 countries. The team consisted of people who actually have been doing the work on the frontline for many years and quickly gained the trust of the local organizations due to their empathetic and compassionate collaboration. They have shared experiences knowing exactly what the most troublesome processes were, while at the same time would simply listen to all the challenges the regions were facing. These were able to be visualized using Process Mining, the relevant questions could be raised, and the answers were derived together in a community approach.

All the improvement topics were transparent across countries while common areas were addressed centrally by the OM4T. This helped the design and implementation of global change requests as well as awareness on a global level of what was really going on in the regions. But while the spirit was positive and the regions were glad to use Process Mining, measuring success was still challenging. It was time for a new KPI, which one of the project team members created: the Digital Fit Rate. The Digital Fit Rate is a very simple calculation. It counts the number of manual activities and divides this by the number of sales order items. While the automation rate is centered on the automation of activities and the rework rate around the manual steps around items with change, the Digital Fit Rate practically addressed both metrics in a single KPI. It was a true KPI, because the regions loved its simplicity and flexibility. The KPI can be broken down to any dimensions such as region, material, or customer (or all at once) while it provided also a global view of status and progression. In Fig. 9.2 you will see an example of how it looks like in reality. As you can see in the example, the KPI and metrics are measured along the dimension of "country," This is a one-click global overview of our executive management and the regional executive management have a continuous feedback of their own ranking on a global level.

It was a rare but welcomed situation, when during the running year targets for the Digital Fit Rate were defined and communicated, countries saw the benefits of such on-the-fly decisions and accepted the new KPI. All other KPIs and metrics that were recorded before became irrelevant immediately, regardless of whether they have been used for over a decade already. Overnight, the Digital Fit Rate became the one and only KPI that mattered in the realm of O2C.

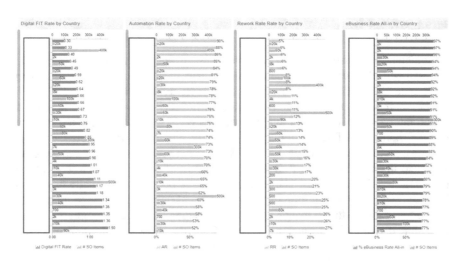

Fig. 9.2 Digital Fit Rate and additional metrics by country

Impact

Using out-of-the-box reports from the Process Mining tool, the results included an increase in automation by 24% and a reduction in manual rework by 11%. In the relevant scope, that meant a reduction of over 10 million manual activities within the first year. Adding our own KPI Digital Fit Rate, we have been able to reduce the KPI by roughly 1.0, which translates to one less touch on average on a global level. The specific levers included further automation of order entry by increasing the efficiency of electronic orders to be flawless e2e from customer to regional company or increasing the automation of back-to-back POs within the Siemens regional company to distribution centers. On the other hand, for some countries new ways of automated billing processes were introduced by sharing knowledge and experience from other countries. While certainly some "low-hanging fruits" were apparent in the very first year, the success has continued until today, 3 years in. Gone are the days of process mapping workshops or initiatives about standardization set as a tone from the top. By the continuous collaboration between headquarters and regions and between regions on all different functional levels from order management, IT, sales, or process consultants, standardization has become a mere positive side effect of furthering automation and decreasing rework.

Altogether, these specific improvements that led to fewer manual tasks also created the confidence to trigger further activities such as the consolidation of order management organizations into order management hubs that would serve Siemens customers across different countries. This idea had been discussed various times over many years, yet a decision for such customer facing activity never gained enough traction. Until OM4T, Process Mining and the Digital Fit Rate came along: within 24 months, three order management hubs have been created in Portugal, Bulgaria, and Poland and around 20 countries are being served from these locations today. This decision was not made due to the availability of Process Mining, but it accelerated the decision and implementation as Process Mining and the Digital Fit Rate were used extensively in the transition project to assure that process performance were not impacted by the decision of organizational change. We did not even follow a standardized approach as every hub had their own journey ranging from Lift-Drop-Change or Change-Lift-Drop. Standardization thus did not mean a standardization of output, but a standardization of positive mindset and direction—made visible for all to see by the use of the Digital Fit Rate.

And now, the Digital Fit Rate has become an integral part of process excellence in every region and the tremendous improvements on a global level continue to new levels of process excellence. The Digital Fit Rate continues to decrease, which means that the amount of human interventions in the process is decreasing. While at the beginning not even a handful of countries were below a Digital Fit Rate of 1.0, in 2019 more than half of our countries are already below 1.0 and the "virtual barrier" of 0.3 has now been surpassed by the first countries, too!

Technology

While clearly our journey using Process Mining was a human-centric one, techno-logical aspects have also been essential in helping hyperscale the impact at such speed. To summarize, there have been four different design principles which were key in the scalability and adoption of the Process Mining technology around the globe.

• No mapping, translation, or conversion of data

Arguably the most important factor for a fast global rollout was the decision never to map, or converse data in the frontend, if the backend did not yield the expected structure. To give a concrete example: If in the backend there is a field called profit center and within Siemens profit centers have to follow a certain structure like being ten alphanumeric characters long, then the value of "PCS System software driver" which is a product description would not make much sense in the field "profit center." Due to "historical reasons," some of the standard ERP fields were "misused" for other purposes and the actual profit center was now maintained in a different field with an obviously different field name. Often the request was: please map the field from the backend to the field in the frontend. However, this request was always rejected due to the following reasons: do not create additional complexity due to a single country's request and to encourage the country to correct data at the source. A positive side effect was the ignition of countless master data cleanups in the source systems (backend) because everyone wanted to enjoy the benefits of using Process Mining (frontend).

• Self-defined activities

One of the most common questions related to the data model were: how did you define the activities in the Process Mining tool? Actually, no activities were defined in the Process Mining tool at all. The name of any activity came from the source ERP system and was retained through to the Process Mining frontend. That meant that it was very possible that at one time the Process Mining tool shows 83 activities and in the next year the number of available activities went up or down. If the source had an activity with a cryptic name like "Act_Desc123%!5_", then we would have honored this description in the Process Mining tool as it is. However, the standardized approach proved that Pareto principle was upheld and "odd balls" occurred infre-quently. Actually, with only 7–8 activities over 90% of all activity occurrences were already in scope—while over 80 different activities overall happened.

• No filtering of data in the backend

One of the most challenging IT requirements put to our project was to only load relevant data, which made sense. Why load data that would never be used? However, how to know what kind of data will really be used? Therefore it was imperative for the success of the project to start small but not limit the creativity of the people and potential of data. The decision was made to load all available data, but for a limited period of time, like a month or so. We ultimately knew we wanted to have 12 months and more, but before we would load the necessary scope it was more important to see what was available in the frontend at all and what made sense. Of course it was imperative to have knowledgeable business users who would ultimately use the system themselves to make this important decision. This meant, in reality, that the project team brought together a wide selection of regions to make this decision together. The fear that this inclusion would slow down the process did not materialize. In fact, the decision process was much quicker than in other comparable projects and changes were often done in days and weeks instead of months and years.

- Process activity focuses on the context

When there is an activity like "Change Price," it is very tempting to nuance this activity by its metadata of the price itself and thus further activities like "Increase Price" or "Decrease Price" are defined. Or instead of "Change Status" the differences of status like "Status: in progress" or "Status: complete" are defined. While this approach might work very well for single system use cases, it would have not worked for Siemens' global O2C use case where so many different systems are connected. The result would have been a classical "different terminology same meaning, same terminology different meaning and neither" situation. Thus, it was based on the context, which was the same for everyone, and not on the content, which would have been specific for every single one. When everyone "sacrificed a little," everyone would gain ultimately than if one had to sacrifice a lot with everyone gaining nearly nothing.

Lessons Learned

The major lesson learned was the sheer power of the frontline experience and collaboration mode of the project that led to a new way of working together with the regions and the small HQ team. The HQ team was comprised exclusively of members who have previously worked in the regions before and thus were quickly accepted by any region due to their empathetic approach and understanding of business processes. They have not only heard about their problems—they actually lived them day by day in the past. Furthermore, the team connected all regions when relevant shared challenges appeared and change was then implemented in a much more rapid way than ever before. A positive challenge between the regions was initiated where country A wanted to surpass country B to have the best Digital Fit Rate. The results of each country were actually shared monthly so that countries could connect with each other depending on maturity levels in automation. The team even sourced a person from the region to join the HQ team during the project to scale the regional champion to become a global one.

Another major lesson learned was that reporting use cases was an extreme risk for project failure. It is very tempting to give in to the power of analysis paralysis. But this is not how change is initiated. People who are living the processes every day need to be able to use Process Mining and understand what value it can bring to each of them. Many times I have seen how other projects used Process Mining as a top-down KPI-driven reporting tool. While Process Mining is an incredible technological progress, it will always only be an enabler for change. Process discovery and analysis have been improved tremendously to identify potential levers, but the actual improvements don't happen in the Process Mining tool but the ERP systems or in the organizational setup with the people, to just name a few. There were correlations where organizations that used Process Mining for reporting purposes only, focusing on the sixth digit behind the comma, had also the worst improvement rate in comparison to other countries.

Countries that used Process Mining to identify specific topics and implemented those measures and then returned to Process Mining to gather the evidence that improvements were visible and then repeated this process were far more successful in improving the Digital Fit Rate. By all means, that does not mean that reporting is not important, but it means that in order to drive change on a global scale that sticks in the organizations, reporting of shared successes that actually have been achieved made more sense then to start with reporting, which allows to show the status and discuss what the status is versus what it should be.

Outlook

Process Mining can be used in any domains and industries and we probably will see Process Mining being used on an individual level in the consumer area as well, like the digital twin of one. However, Process Mining will continue to emerge into such pressing matters as sustainability as well. New business models will arise around the reduction of waste and other inefficiencies of resources, be it energy, CO_2 emission or plastic usage. Just as companies like the Plastic Bank is turning plastic into cash for the poor, Process Mining can help such companies to monitor the flow of plastic cradle to cradle in order to better understand where in the value chain waste can be reduced, e.g., by improving recycling infrastructures, less consumption by alternative products, or the usage of recycled plastic. I personally see Process Mining as an independent and universal companion to create data-driven but process-centric transparency on any use case, yet especially in the still underdeveloped area of our most pressing questions around sustainability. It would be my personal wish to see more and more concrete use cases around efficiency of monetary activities from donations or grants, usage of natural resources or consumption of waste, or achievement of sustainability development goals.

Furthermore, from a technological point of view, we will probably see further convergence of Process Mining with other technologies such as machine learning, RPA, or graph databases as well. Visualizing processes to make them more communicable across a set of stakeholders to then identify improvement measures is one thing, but to predict future processes and outcomes or to not only recommend but also execute the next best action are just two examples where Process Mining will need to go. With the vast experience and knowledge of Process Mining within Siemens across many different use cases, we are not only ready but also will drive the necessary actions toward the fulfillment of this potential.

Links

Metrics of Success. Sharing the journey to a wider audience is also an integral part as it was during the program internally: https://www.youtube.com/watch?v=RkiggZTRbog

Digital Fit Rate—the first chapter of our ongoing digitalization story: https://www.linkedin.com/pulse/digital-fit-rate-first-chapter-our-ongoing-story-gia-thi-nguyen/

From Complexity to Competitive Advantage—Cutting the Gordian Knot (2018 edition): https://www.linkedin.com/pulse/from-complexity-competitive-advantage-cutting-gordian-nguyen/

Uber: Process Mining to Optimize Customer Experience and Business Performance

10

Martin Rowlson

Abstract

Process Mining has allowed Uber's Customer Support teams to uncover insights across their processes that touch more than 700 cities across 65 countries on 6 continents. This capability allows Uber to understand variation in customer support and target large-scale multimillion dollar efficiency gains through process harmonization and increased customer satisfaction though global process benchmarking. Internally, Uber has used the power of Process Mining to help foster a culture of continuous improvement by providing a deeper level of business process visibility.

Challenge

Upon joining Uber back in 2016 as the sole lead for Process Excellence in Uber's Support Operation, it became immediately obvious to me that Uber's rapid growth had resulted in high process variation leading to lack of consistency in how we handled customer support issues. This lack of consistency created additional process waste and ultimately increased cost in agent handling time and in some situations created poor customer experiences due to a variation in the customer outcomes. If we had the ability to quickly identify all our processes and uncover unnecessary variations we would ultimately handle customer issues faster with less effort, and in some cases with automation requiring no effort at all. On top of this huge variation challenge it was very difficult to obtain reliable data allowing anyone to understand how the processes are performed across our large agent footprint.

The way to really understand how process is performed was to "go see." By observing a sample of agents performing a given process, we aimed to infer process

M. Rowlson (✉)
Uber, Amsterdam, Netherlands
e-mail: m.rowlson@uber.com

© Springer Nature Switzerland AG 2020
L. Reinkemeyer (ed.), *Process Mining in Action*,
https://doi.org/10.1007/978-3-030-40172-6_10

improvement opportunities. Not only was this approach not scalable, it was not statistically significant. Obtaining data on a small set of agents would never provide the reliable data to support large-scale change. This lack of data that supported recommendations often led to subjective discussions about data integrity and why a given process shouldn't change. To compound this challenge we also had constraints in obtaining "accurate" data since this capability was limited to specialized analytics teams. When this constraint exists, one must be either be very influential or you must simply wait until the analytics team have the bandwidth for the ad hoc data requests. In a company with so much scale, it was imperative for the Process Excellence function to obtain data at a higher frequency in a larger volume across all cities, countries, subregions, and mega-regions.

Use Case

Community Operations support footprint spans many thousands of agents with large-scale operations in several countries covering the companies need to provide support to customers across the globe. The use case deployed for Process Mining in Uber is to identify opportunities to improve the e2e process of customer support contact handling with the goal of creating efficient (faster response) and harmonized processes that lead to fewer response errors or rework loops. All with the aim of providing a consistent high-quality service to all of our customers and partners 100% of the time.

What does the customer experience look like and where are we focusing?

In order to illustrate our use case with additional clarity, it's important to explain the customer journey and where we are applying Process Mining. Customer support within Uber deals with a vast array of customer questions across pre-trip (onboarding, setting up accounts, etc.), on-trip (Live Eats orders, trip safety, etc.) and post-trip (fare issues, navigation). In all scenarios it's imperative that we provide a fast and correct outcome. For this to happen we must have full transparency of our operational processes. Traditional reporting methods, such as scorecards, only go so deep in terms of uncovering process inefficiencies. They typically notify someone when a key metric degrades. Process Mining allows us to uncover unknown process wastes that scorecard metrics wouldn't uncover. Such issues like inefficient contact handling due to multiple hand-offs and typical LEAN transport wastes.

In terms of the implementation method used for Process Mining, we chose to implement our solution with a direct connection to our data warehouse, meaning we could push the required data on a daily basis and not worry about having to pull data into our tool when required. This approach meant that anyone in the organization could have access to the same data and insights without having Process Mining as a specialized capability within the Process Excellence team (making powerful insights more accessible enabling a continuous improvement culture).

In order to gain the maximum benefit out of Process Mining, we had to think about an internal support function that would ensure (1) the data pipelined into the tool didn't break and (2) the dashboards we designed were functional and met the needs of the end users. This saw the creation of a small team of data scientists whose sole focus was ensuring the tool performed as needed and the end users were trained with the required skills to process mine. This team served as points of contact for other business departments to help them scale out the capability locally; essentially we created a CoE for Process Mining.

Impact

Looking back at the challenges articulated in the earlier sections regarding how we gain deep process insights at speed across our entire support process landscape, we have been able to eliminate almost all of these challenges. We are able to analyze process data with such speed that allows us to accelerate the realization of process improvement opportunities. With this capability it's also possible for us to identify inefficiencies we hadn't thought to look for—e.g., having visibility of how process is performed across all of our support interactions uncovers small quick wins and often hidden inefficiencies. We are primarily using Process Mining to improve customer satisfaction and can track the impact in metrics such as Average Handling Time (AHT). In regards to the latter, we are now able to compare like for like processes between agents, sites, cities, etc., and align to best practice or even better automate the process through RPA.

With the implementation of Process Mining, we have quickly identified over $20M in efficiency gains through handling time improvements.

Technology

We needed a solution that had the capability of handling billions of rows of data with refresh rates in the seconds and not minutes, along with an intuitive flexible user experience. With an intelligent business cloud solution we were able to directly push the volume of data we needed and connect it up directly into the back end of the tool. This just left the data visualization part to the local team, where anyone in the business can interact with the tool to process mine pre-built analyses and a select quantity of people can build analysis dashboards or connect data pipelines.

Lessons Learned

The implementation of Process Mining within Uber's community operations team was largely a success given the huge efficiency gains and the small amount of resources involved in implementation (team of four comprising of myself, Uber IT engineer, and two data scientists from our vendor). Some of the lessons learned came from understanding the audience for Process Mining. Originally, I thought this would be our analytics and insights community. It turned out to be our less analytics savvy employees who could now perform deep dive analytics without requesting a large amount of SQL queries from analytics staff. We now have Process Mining available and used across all departments in Customer Support and we are exploring use cases in Financial Operations.

Looking back at the Process Mining dashboards my team and I designed at the beginning, they were not as user friendly as we originally thought. This became more obvious when performing demonstrations that my team and I found relatively easy due to our experience, yet our audience lost the ability to follow what we were doing. Once this lightbulb went off it became our goal to create "beautiful dashboards" that were both visually appealing and easy to follow. The latter came by the way of narrowing our focus from broad Process Mining across all metrics to the creation of dashboards designed to focus solely on specific key metrics such as AHT or Customer Satisfaction CSAT rather than one dashboard combining everything.

As with all new tools or capabilities, the acceptance within the business was mixed. We had the typical "it's another tool" syndrome, but that was also balanced with a level of excitement within the non-SQL-proficient groups about what analysis is now possible. It was also important for me and my team to clearly articulate where in the analytics landscape Process Mining sat. We didn't want to get into nonproductive conversations about Process Mining replacing existing reporting tools such as Tableau. For us, the Process Mining layer sits cleanly below current reporting tools where we need to go deeper into the process to uncover the insights and at the bottom-up level where we mine for issues that reports can't identify.

Outlook

Looking beyond at the future of Process Mining, I envisage a place with less requirement on the ability to "Process Mine" and more proactive by leveraging machine learning to drive predictive analytics. With any system that consumes billions of rows of data, there is an obvious opportunity for the tool to start to predict outcomes by reading trends in the data models. I can also picture the landscape of analytics teams changing from ad hoc data analysis and dashboard/report maintenance to powerful insights teams who drive waves for process improvement insights. The dynamics of a weekly leadership meeting is also subject to change from looking at static dashboards showing upward and downward trends in data to a dynamic meeting that would allow people to come to meetings with known reasons for shifts in key metrics or even one step further where we can perform metric deep dive live within business reviews to uncover root causes. This would lead to a massive improvement in driving insightful and actionable meetings.

Process Mining will also continue to complement current automation (RPA) programs. In order to automate the correct processes one must have a good understanding of the processes in question. Process Mining allows organizations to truly understand how their processes are performing versus how they think their processes are performing. This level of knowledge perfectly complements current automation efforts.

The long-term vision I foresee is that Process Mining will evolve and merge with natural language and machine learning processes to realize the potential for over 95% of support contacts across any industry being handled digitally, leaving only the critical contacts to be solved by humans.

BMW: Process Mining @ Production

11

Bringing Innovation to Production Processes and Beyond

Patrick Lechner

Abstract

It has been about 3 years since BMW Group first started using Process Mining—besides several other fields—in an area where probably no other company had used it in such depth and with such an impact before: in manufacturing/production. When Nicolas Größlein and I introduced Process Mining at BMW Group with the great support of our former CIO, Klaus Straub, and our Vice President Connected Vehicle, Digital Backend and Big Data, Kai Demtröder, we were not driven by the desire to be particularly innovative or creative. Our main driver was to ensure world class production and the best possible quality of our cars for our customers. Because premium products require premium production processes!

But is Process Mining at production really a new driver for innovation that can bring production processes to the next level? Or is it just a hype, a buzzword that will be replaced by the next one pretty soon? For BMW Group it has turned into a game changer, as it is shown in this use case.

Challenge

The automotive industry is currently—probably more than any other industry—undergoing massive change. In the next 10 years we're likely to face more changes than in the last 50 years. These range from digitization to electrification, autonomous driving, and alternative traffic concepts in metropolitan areas. New competitors, e.g., from the USA and China, are massively increasing the pressure on the established German manufacturers and drive them to act.

P. Lechner (✉)
Process Mining & Robotic Process Automation (RPA), BMW Group, Munich, Germany
e-mail: Patrick.lechner@bmw.com

© Springer Nature Switzerland AG 2020
L. Reinkemeyer (ed.), *Process Mining in Action*,
https://doi.org/10.1007/978-3-030-40172-6_11

Process Mining is an increasingly important tool for the BMW Group in virtually all areas of the company in order to be prepared for this. The acronym VUCA describes the challenges that our ever-changing world holds for companies:

- Volatility/Variability
- Uncertainty
- Complexity
- Ambiguity

No one can say with certainty how markets, competitors, and technologies will change in the next few years and how we can best position ourselves. But what we do know is that they are changing and that we need to respond quickly to these changes in order to stay competitive. While we were able to adequately address constant challenges with established strategies, structures, and processes, in our modern world we increasingly need the ability to quickly adapt our strategies, structures, and processes to the actual conditions (with the same product quality). This new way of thinking and working can be summarized under the term "agility," The most agile companies in the industry are 2.7 times more successful than their competitors in a 10-year comparison.[1]

Process Mining is a key tool for the BMW Group to meet these new challenges and make our processes fit for the future. It helps to improve agility through four main factors:

1. Agility through transparency
2. Agility through standardizing
3. Agility through speed
4. Agility through quality

BMW has always been a driver for innovation for more than 100 years:

- In 1917, BMW developed the first aircraft engine with aluminum pistons (BMW IIIA).
- In 1935, BMW introduced the world's first hydraulically damped telescopic forks.
- In 1954, the world's first aluminum V8 engine for series production entered the market.
- In 1973, BMW was the first company in the automotive sector to appoint an environmental officer.
- In 1988, BMW (in cooperation with FAG and Hella) developed the first antilock braking system for motorcycles.

---

[1]European Agile Performer Index—determined by goetzpartners and the NEOMA Business School.

- In 1991, the BMW E1 became the first car of the modern age designed from the ground up as an electric vehicle.

Those and many more "firsts" show that innovation has always been in the heart of BMW. Only through innovation can a company in the automotive sector be successful over such a long time! Next to this ability of being innovative, being agile will—in my opinion—be a second crucial factor for success in the future. Therefore, Process Mining became a game changer for us about 100 years after BMW was first founded in 1916.

Use Case

The initial motivation to use Process Mining for our production processes in 2017 was the introduction of a new, highly innovative paint shop in one of our plants. As with most new technologies, this introduction did not happen without friction and the experts in the paint shop tried to identify the root causes for these frictions (e.g., errors in the paint for certain colors, long runners, necessity for rework).

This was mostly done in the old-fashioned way: through observations, learnings from years of experience, and trial and error. Although a significant amount of data had been collected from each production step (through sensors and manually), the evaluation of this data was often complicated and time consuming. Analytics had been used for various specific production challenges in the past. However, data-based process monitoring and process improvements had not yet been in place to a significant extent.

And this is where Process Mining became a game changer!

To be able to apply Process Mining in our production processes, relevant sensor data needed to be extracted. Therefore it is of great benefit that BMW Group is using (almost) the same IT system in (almost) all plants to collect and store this data.

First of all our paint shop experts have been quite critical:

- Why do those IT colleagues think they can solve problems that are hard to tackle for the most experienced and skilled experts in the field?
- Can the data that they had been collecting for years really provide extra insights and drive improvements?
- Or will there be incorrect conclusions that only cause extra work or even make the issues worse?

While collecting the data is relatively simple, if you have a good IT landscape in place at a plant, evaluating the data is more difficult:

- While standard finance or procurement processes may have a few dozen distinct activities, in modern paint shops hundreds of sensors collect process data that needs to be evaluated to fully understand the processes: Looking at all production sub processes, this number can easily be more than 1000.

- Working calendars can be highly complex: we need to consider complex shift models, machine maintenance windows, trial runs, etc.
- For many use cases data needs to be evaluated and analyzed with a high frequency. In order to get most out of the data, near-real-time analyses can be essential. Only then can we react quickly and act to avoid/minimize production problems.
- Data quality is essential. Incorrect measurements of the sensors should ideally be identified before the analysis. This might require expert knowledge and complex filter criteria.

Impact and Technology

As soon as these challenges have been mastered, Process Mining can bring massive possibilities for production responsibles:

- We can visualize the production processes as they really happen and not as they've once been planned. Complexity and interdepartmental cross overs can be understood in detail and misunderstandings can thereby be avoided.
- By creating a digital twin of our production plants we get more agile, since we can identify issues faster and react immediately.
- This effect can be amplified by near-real-time connections to the source systems. Based on this, we can use Process Mining as a monitoring tool in production that allows data-based, fast decisions. These can help to reduce rework, improve the quality of our products, and reduce production costs.
- We can compare the "As Is" process with the originally planned process. Thereby we can identify unplanned deviations as well as requirements to modify our production steps.
- Bottlenecks can be identified quickly. We can see which process changes really make sense, e.g., if a parallel queue of certain production steps can help to reduce throughput times or if these changes would be at the wrong part of the process to achieve this.
- By adding quality data to the analysis, we can perform a data-based root cause analysis. Certain errors in the production can be linked directly to the relevant process steps and these can then be addressed and improved.
- We can analyze the quality and relevance of process KPIs. Steering by KPIs can help improve performance. However, it can also lead to "creative" ways to achieve KPIs, that are not value adding or in some cases even negative for the overall output. By looking at the process in detail instead of single KPIs we can identify unwanted effects and in some cases modify KPIs in ways that really optimize the output.
- By adding cost figures to the analysis, we can analyze the exact production and energy costs per car. In the example of the paint processes this helps us to identify

colors for which the production costs are higher than planned. By doing so, we can either search for possibilities to reduce these costs, not offer these colors anymore, or modify costs for the customers for certain colors.

• By introducing Process Mining at several plants, we can compare the process quality and performance in these different plants. Although there are of course many plant-specific differences, certain comparisons are possible. Benchmarking and learning from other plants help to achieve best results in all plants!

After a detailed market analysis, BMW Group decided to use a leading provider as our main tool for Process Mining in 2017. Due to its usability, this choice of software enables us to scale Process Mining fast and successfully and to give our business units the ability to develop their own analyses. Next to using commercial products as a basis, close collaborations with universities and other research institutes as well as own developments help us to stay ahead of our competitors in the area of Process Mining. Use-case-specific gaps in the software can therefore be closed quicker and special requirements can be implemented perfectly.

Since easy and fast access to data is crucial for regular process improvements, the BMW data lake (on premise and on the cloud) is one of our most important data sources. For near-real-time analyses we also use direct access to systems, respectively access to mirrors of these systems.

Lessons Learned

• *Transparency can cause pain and adversities:* Process Mining should not be used for naming and blaming, but to improve our processes. No one should be punished for non-optimal processes in the past, but everyone should get all necessary support to achieve better processes in the future.

• *Introducing new technologies is not easy:* For Process Mining it helps to present the functionality in live demos based on real data of the business unit. By concrete examples the power of Process Mining can be communicated best.

• *Naming (quantitative) business value for analytics is not always popular and easy.*

   This has several reasons:

   – Visualizations and monitoring can be extremely important for the business. However, it is hard to quantify its business value.

      How much time can be saved based on these visualizations?

      How can we improve quality or reduce costs just by seeing the facts?

   – We have to ensure that business is not punished for naming and delivering business value. Money or resources saved should be reinvested in the business units, which achieved it, for future investments and improvements. Only by giving good incentives can we achieve the best result overall for our companies.

   – Analytics itself does not normally create business value on its own. Only if certain measures result out of the analysis, we can realize the identified potentials. These measures can be organizational changes, changes in the relevant IT systems, etc. But which part of the achieved business value is due to analytics and which part is due to performing those measures? While this does not matter for the company overall, it has a huge effect on establishing data analytics methods in a company.

- *Scaling and rollout is not easy:* We need to have sufficient support inside the plants to introduce Process Mining fast and to achieve the best results. Therefore we need to ensure this support and we need to train business units before rolling out Process Mining in a plant. With the help of key users, we can then scale up quickly and support the plants within a short time.
- *One size does not fit all:* Even between different plants or between different technologies, we need to have several support models. While some business units very much enjoy developing their own dashboards, others would like to have a service provider to act as a dashboard factory for them. Ideally we have the possibility to offer both. Therefore a Process Mining Center of Excellence can act either as:
  - Consulting Team
  - Self-service enabler
  - Platform management service only

Which one works best for a certain business unit needs to be decided together with the business responsibles.

- *The Process Mining world is changing fast:* New tools and technologies flood the market every month and we need to monitor those closely. We need to stay informed and make wise choices about which new technologies to use and which ones to ignore.
- *We need to establish a strong user community:* Creating Process Mining analyses in areas where no other company had been using it before means that we have to create everything from scratch. Exchange between all users is therefore essential in order to get the most out of these analyses.

We get most out of process improvements if we improve the processes along the whole value chain. If bottlenecks remain, subprocesses of this chain and optimizations in other subprocesses might only have a limited effect. The value chain of a car manufacturer can be distinguished in its Primary Activities and Support Activities. The *Primary Activities* include the following (in chronological/ procedural order):

1. Inbound Logistics (raw materials handling and warehousing)
2. Operations (machining, assembling, testing products)
3. Outbound Logistics (warehousing and distribution of finished products)
4. Marketing and Sales (advertising, promoting, pricing, channel relations)
5. Service (installation, repair parts)

Parallel to these Primary Activities several Support Activities help car manufacturers to run smoothly and enable this Primary Value chain. The *Support Activities* include:

- Firm Infrastructure (general management, finance, accounting, strategic planning)
- HR Management (recruiting, training, development)
- Technology development (research and development, product and process improvement)
- Procurement (purchasing of raw materials, machines, supplies)

While Process Mining has already been established for the Support Activities in many companies, we've mainly focused on the Primary Activities at BMW so far.

Among our analyzed Primary Activity Processes (next to the previously described production processes) and the questions asked are the following:

- *Warehouse Management:* How can we minimize storage times?
  How are goods moved around the company? How can we minimize these times?
- *Development Processes:* How can we reduce our development cycles?
  How can simplified platforms and less variants help to get more agile?
  Where can product tests be simplified or optimized?
- *Customer Experience:* How do customers use our products?
  Which functionalities are not really used a lot and could be removed to minimize development costs without reducing customer experience?
- Where do customers struggle with the products' functionalities?
- *Change Management:* How are changes that come from development adjustments, new suppliers, legal regulations, etc., moved into production?
  How can communication channels be optimized between these different departments?
  What are the right next steps for the employees in the plants to set the changes live as soon as possible?
- *Car Distribution:* How can we minimize the time from the plant to the customer?
  Where are the bottlenecks or regular issues in the distribution?
  How can we inform our customers better about the progress and when the car will arrive?
- *Aftersales:* How can we offer our customers the best possible aftersales experience?
  What is the best way to retain customers through aftersales activities?
  How can we optimize the network of suppliers for spare parts?

Outlook

We have come a long, long way since first starting Process Mining in 2017. But there is a lot more to come.

And there are many challenges and chances ahead of us:

- *Stronger focus on support processes:* While we're doing really great already at using Process Mining for the Primary Activity processes, there is still a lot of potential for support processes, as they get increasing focus in our analyses.
- *Process monitoring vs. process improvements:* In a first step many business units focus mainly on visualizing and monitoring their processes. In order to get most out of Process Mining; however, it is crucial to also use it for process improvements. Getting there is not always easy, since process improvements are not always part of a business unit's targets.
- *Further scaling:* Process Mining as well as other analytics topics will become more and more relevant in the future. Therefore it is important to build up the relevant skills. Those are still quite rare on the market and we therefore need to ensure that we stay attractive as an employer. At the same time we need to ensure that Process Mining can be used more and more as a self-service tool. Only then will we be able to scale fast and successfully.
- *Process Mining and RPA:* The past months have shown clearly that Process Mining and RPA are moving closer together. This makes a lot of sense since one of the most common ways to improve process is the automation of parts of the process. RPA therefore allows to directly get higher business value out of our analyses.
- *Process Mining and further artificial intelligence analysis:* In many use cases AI analyses can help us to dive even deeper into the data. AI offers possibilities, e.g., in the area of predictive Process Mining.
- *Digital transformation:* Due to the expected, massive changes in the automotive sector, numerous optimization programs will most probably be launched in the coming years. Experience however shows that transformation programs are often not very successful due to process complexity, lack of transparency, and inadequate monitoring options. Process Mining will help us to close precisely these gaps and optimally support transformation and agility. Therefore Process Mining will stay a game changer in the future—moving from a process visualization and process improvement tool to becoming a digital transformation tool.

Process Mining will therefore play an important part in ensuring excellence in our company's processes. It will help to make BMW Group ready for the future—probably for the next 100 years of innovation!

Links and Further Reading

Kai Demtröder, Dr. Patrick Lechner: "Agilität durch Process Mining", Manager
    Magazin, Manager Wissen, June 2019

Siemens: Process Mining for Operational Efficiency in Purchase2Pay

Khaled El-Wafi

Abstract

Purchase-to-Pay (P2P) Process Mining has the objective to visualize process flows, identify process weaknesses, and support process improvements. It allows to monitor and manage any process in complex, global organizations in an unprecedented form and efficiency. This includes:

- Visualization of P2P processes based on live data from SAP ERP systems. Time stamps for duration between relevant process steps.
- Identification of P2P process weaknesses, e.g., with low degree of automation or multiple approval steps.
- Support process improvements with immediate review of process adjustments and interactive remediation.

In a nutshell, P2P Process Mining provides the answers to "how can I increase operational efficiency?" and "how can I optimize my working capital by reducing cash out towards external suppliers?" within the P2P process.

Challenge

P2P processes become more important for business as the world is moving faster and as goods and services need to be delivered within a short time in order to maintain the required advantages over competitors. According to Porter's five forces, an intense competition can only lead to an increase in market shares, if the business continuously monitors how to deal with vendors/suppliers, customers, substitutes, and new providers. Therefore, the own competitive situation can only be managed

K. El-Wafi (✉)
Siemens AG, Munich, Germany
e-mail: Khaled.el-wafi@siemens.com

© Springer Nature Switzerland AG 2020
L. Reinkemeyer (ed.), *Process Mining in Action*,
https://doi.org/10.1007/978-3-030-40172-6_12

by determining your competitive level and through the estimation of your profit potential.

An intense competition can lead to a sector growth, new providers might lead to economies of scale, market identity, venture capital need, and political influence, whereas customers have clear expectations for expected volume, pricing, brand identity, quality, performance, and influence on stakeholder decisions.

Suppliers/vendors are being measured from costs, quality, process, and speed perspectives within the P2P process.

- Pricing, PO volume, and invoice volumes are relevant metrics for monetary/cost tracking of supplier performance.
- Supplier delivery reliability and delivery capabilities are relevant metrics for tracking the quality of supplier deliveries, which consist of materials, goods, projects, solutions, and services.
- Activity flow, throughput times between different activities, and avoidance of inefficient activities are related to process transparency and, if well managed, lead to process optimization of the Supply Chain Management (SCM) process.
- Throughput times between different activities, if well managed and automated, lead to an increase in process speed.

For that, it is of outmost importance to start focusing on suppliers/vendors efficiency, as the goods purchased from them represent a high percentage of the overall expenses of the business, reaching in some cases more than half of its revenue. In the following chapters, the term suppliers will be used as part of the SCM core process, which is to be further optimized.

Segregation of Duty Versus e2e Transparency

The P2P process is part of core processes and has to be monitored from both the Purchasing and the Payment perspective. Procurement departments are more focused on the purchasing side, whereas the accounting departments deal with invoices and release/blocking of payments. In the last decade, the P2P process was more split between the first P (Procurement) and the second P (Payment) for segregation of duty reasons with the clear goal to avoid fraudulent behavior in internal organizations with regards to ordering and receiving goods as well as paying the invoice without a clear distinction on who is approving what process step.

Thanks to the new digitalization capabilities it is nowadays possible to identify fraudulent behavior within the P2P process and therefore an artificial separation of the first and second P is no longer needed. Such a type of separation might even be considered as contraproductive for the health of a business. With Process Mining, business is even encouraged to cover an e2e process, as it is more efficient to consider all facets of the process rather than to focus on a fragment of the overall case. In a global world, customers can compare goods and services from suppliers within short laps of time, which leads to higher demand of quality and faster delivery.

Cornerstones and Definitions for P2P Process Mining Cases

The following cornerstones and definitions are required to understand the P2P use case.

Supplier
A supplier is to be differentiated between internal and external. Internal suppliers are part of the value chain and are relevant for internal charging, which is called intercompany clearing. For external suppliers, additional rules apply with regards to PO approval and release as well as the charging/payment process. An external supplier in general is tracked by the business from different angles to enable it to produce goods or services for delivering them to the purchasing organization.

Material
Suppliers deliver all sorts of materials to the procurement and logistics departments of the business with the aim to produce goods and services that are sold to external customers. Therefore, the availability of a comprehensive set of report details based on material details shall not be underestimated. The included information might cover, among others, forecast data, budget data, and technical details on a material level in various dimensions. This would include filters according to different dimensions such as business unit, plant, currency materials, material areas, commodities, etc.

The objectives of stringent material reporting are to provide detailed material data, including procurement info record, material master data, and eventually technical specification. Detailed source volume, quantity, budget, and forecast data on material level are of a big advantage for the business as it saves time and increases efficiency in the daily work of the employees working in the fields of procurement and logistics.

Key Performance Indicators Versus Business Volume Indicators
P2P KPIs and Business Volume (BVI) Indicators are used to monitor the health of the supply chain organization. A KPI, e.g., "Automation rate" calculates either a rate in [%] percentage or a ratio in [absolute] value. A BVI, e.g., "# of POs" is a helping metric that displays an absolute figure with no possibility to compare it within an organization or across organizations. These metrics combined with Process Mining enables the management to derive the right measures from operational efficiency and capital costs perspective.

P2P KPIs/BVIs can be categorized into:
- *Risk indicators of a supplier*, e.g., geo risks or political risks for quality and allocation management. Material undersupply cases based on monopoly codes

per material enable identifying suppliers that need to be further evaluated, qualified, and if needed further developed or phased out.

- *Material indicators*, e.g., the amount of different materials and commodities per supplier, quantity and volume per material in local currency, risk ratio of material undersupply, second source ratio.
- *Purchasing supplier key performance indicators*, e.g., supplier delivery capability and reliability, supplier failure rate, automation rate, robot process automation rate, digital supplier fit rate, rework rate, electronic interchange rate (e.g., orders, order change, order response, dispatch advice, delivery forecast, inventory report message types), throughput time in days, hours, or minutes from one activity to the other (e.g., from PO sent to record goods receipt), percentage happy process path, incoterms discrepancies.
- *Purchasing supplier business volume indicators*, e.g., # of payment terms in master data or orders, # of process variants, # of POs, # of PO items, PO volume, goods receipt volume.
- *Accounting supplier key performance indicators*, e.g., automation rate, robot process automation rate, rework rate, electronic interchange rate (e.g., invoice, self-billing-invoice message types), throughput time in days, hours, or minutes from one activity to the other (e.g., from record goods receipt to payment of invoices), percentage happy process path.
- *Accounting supplier business volume indicators*, e.g., # of payment terms in invoices, # of process variants, # of invoice documents, invoice volume.

One of the most underestimated data are the coordinates of suppliers that must be centrally stored in a corporate Master Data. The required information contains address/country supplier information, latitude and longitude coordinates, phone/e-mail information, and the internal supplier number, including the respective global DUNS of a supplier.

Aggregation Levels of P2P Data

Whenever you start creating the P2P data model for your business, it should be clearly defined which filtering criteria shall be used. It is recommended to implement the aggregation of the data with the help of calculation views on the backend development system using Structured Query Language (SQL) packaged stored procedures, which can be transported from development over quality to the productive system of the business owned central data lake. The central data lake shall either be an "on premise" installation or a "cloud solution hosting." This approach reduces performance issues on the frontend business intelligence Process Mining platform.

Filtering criteria allow to select and analyze the process according to specific conditions, e.g., choosing a specific company code and purchasing organization.

- PO year-month: year and month of the PO creation date
- Legal entity: accounting unit number
- Company code: filter by a specific company code
- Business unit: filter for a specific business unit

- Purchasing organization: filter on a specific purchasing organization
- Vendor type: filter by external or internal vendor type
- PO document type: filter on a specific document type
- PO type: filter on a specific type (PO for direct or SRM for indirect material)
- Deletion indicator: filter on a specific deletion indicator
- PO item value range: filter in a determined value range of PO items
- PO header value range: filter on a specific range of values for a PO header

Use Cases

P2P Process Mining standard process steps are shown in the Fig. 12.1 and it can be further subdivided into "Direct" and "Indirect" material focus areas.

P2P Direct Material Focus Areas and Data Models

From planning side, focus areas have been defined for the P2P process as business levers that drive Process Mining, which helps in breaking them down to use cases in business language. From an analytics perspective, indicators are defined with a textual description and, also with the support of definitions and mockups. A toolset enables the development of a data model or framework which integrates entity-specific characteristics that cover information required by the indicators.

On the acting side, Process Mining interactive dashboards need to be created that are specifically targeted to the selected use cases. Collective sessions to identify optimization potentials and root causes support process analysis. The summary of collective sessions as well as the quantification of the monetary and timely impact of potential process deficiencies lead to the development of measures.

Two types of data models need to be implemented in the backend system of the data lake, one for Purchasing and the other for Accounting views. All you need for Process Mining is to retrieve the digital trace of the different activities of P2P starting with a PR item and ending with a payment of an invoice document.

- *The Purchasing model* starts with the purchasing order item and uses Process Mining graph theory to display the previous and next activities generated after the creation of a PO item. This means that at the time where the PO was generated some of the goods receipts or the invoices might not have been received yet.

Fig. 12.1 Standard P2P process steps

- *The Accounting model* starts with the invoice document and it also uses Process Mining graph theory to display the previous and next activities generated after the creation of an invoice document.

It is recommended to develop the two mentioned data models in one project schema on the data lake backend system by replicating the relevant P2P meta tables and table filters from the respective source systems. SQL Stored Procedures shall be developed, and calculation views shall be created for retrieving digital traces with regards to cases (case IDs), activity mapping and time stamps per activity as well as a sort order per activity. A data scientist is definitely required with the right SQL programming skills to map the raw data into stored procedures.

Based on jointly defined P2P used cases, six focus areas are being considered:

P2P Overview: Direct Material

The model shows an overview of all defined activities (process steps) in the Process Mining process explorer and the distribution of PO Items (purchasing perspective) along common identifiers. It is also possible to analyze how many changes are going through rework activities and what is their throughput time. It provides a variety of drill down information to substantiate or contradict new hypotheses.

Here are some example questions that can be answered using this dashboard:

- For how many PO items has the payment block been removed?
- How many PO items end with their creation?
- For how many PO items is the invoice scanned before a goods receipt is recorded?
- How many PO items are affected by a reverse goods receipt?
- How many PO items end with booking the invoice?
- How many PO items have been paid before goods have been received?
- Which material group was primarily affected in this month?
- What is the rework rate by company code or company type?
- What are the top rework activities?
- What is the subprocess, document type throughput time?
- What is the median throughput time in days?

Figure 12.2 shows a visualization of the different variants of the P2P process.

This use case covers all process variations, including cut-off process instances and nonstandard procurement types. It provides a variety of drill down information to substantiate or contradict new hypotheses.

- e2e transparency on the P2P process from the first request until the invoice gets cleared
- Validation and monitoring of KPIs and BVIs
- Starting point to explore new paths for analysis

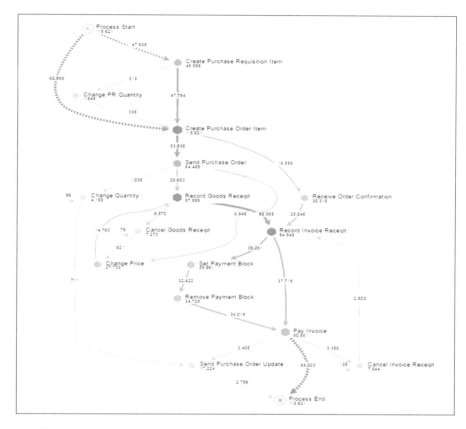

Fig. 12.2   Visualization of the variants of the P2P process complexity

Rework: Direct Material

An example of the rework topic is the fact that price changes in the purchasing process do not only slow down the purchasing process as they are typical rework activities. At the same time, the significant manual effort involved in changing prices also significantly increases process costs. Therefore it is in every company's best interest that the direction of improvement is to minimize the price change ratio as much as possible. Events are defined as rework, if they are not necessary in an "optimal" process execution. Some event types are only treated as rework if they occur more than once for a given process instance.

- Transparency on frequency/clustering of rework
- Identify the impact of rework on process efficiency
- Remedy reasons for unnecessary rework
- Automate not necessarily manual rework events

An activity might be considered as rework on its first occurrence in a case (ex: "change price (ERP)") or only when it happens more than once (ex: "release PR item").

Throughput Times—Direct Material

All combinations of start and end events can be analyzed in the process explorer.

- Transparency on throughput times within purchasing and accounting, respectively
- Remedy root causes for delays in the process
- General streamlining of the purchasing process

Automation and Robot Rate—Direct Material

Too many companies have troubles with the creation of POs, resulting in a loss of time through a high amount of data entry error and manual work. A further effect is the automatic validation against master and transactional data as well as your configuration settings. Another issue is the loss buying leverage and higher costs per transaction, because if a purchaser enters information manually, he or she might miss out on preferred suppliers, pricing, and other terms. Overall automated PO leads to faster processing times, cost savings, and more control.

Incoterms: Direct Material

Incoterms are considered different if any of the combinations in Master, Header, and Item do not match, except for the blank cases of PO Header or PO Item data. In these cases, the comparison is only between the nonblank incoterms and the master data.

Payment Accuracy: Direct Material

Payments are treated as early if the invoice gets cleared more than 2 days prior to the first due date. Payments are treated as late if the invoice gets cleared only after the last due date passed. All other payments are on time.

- Transparency on early, on-time, and late payments
- Remedy root causes for losing cash discount and/or paying late
- Analyze reclaimed cash discount

P2P Indirect Material Focus Areas (SRM)

Supplier Relationship Management (SRM) is a vendor management application that is based on an SAP software platform with the clear goal to order Indirect Material goods, which are usually used in a production process, but not directly traceable to a cost object, e.g., overhead costs. In order to distinguish between direct and indirect material for production purposes, indirect materials are not monetarily significant. Typical process steps of the Indirect Material process for creating and approving a shopping cart is listed in Fig. 12.3.

Fig. 12.3 Indirect material shopping cart process steps

The process times are measured the following way.		
Process Step	From	To
2A	Shopping-Cart Send into Approval Date	Final Approval of Shopping-Cart by Procurement
2B	Final Approval of Shopping-Cart by Procurement	Final Approval of Shopping-Cart by Expert
3A	Final Approval of Shopping-Cart	Creation of Purchase Order
3B	Creation of Purchase Order	Purchase Order sent into approval
3C	First Procurement Approver Receives Workitem	Last Procurement Approver Submits Workitem
3D	First Expert Approver Receives Workitem	Last Expert Approver Submits Workitem
3E	Requester receives workitems	Requester Submits Workitem
4	First Management Approver Receives Workitem	Last Management Approver submits workitem
4P	First Procurement Release Approver receives Workitem	Last Procurement Release Approver submits Workitem

As part of the procurement process, the shopping cart is checked for completeness and correctness. For instance, the procurement commodity code is checked. During the completion stage, the PO is created by procurement. Order approval is done by the relevant managers. Once approval has been obtained, the order is sent to the relevant supplier. Goods receipt can be posted both in the SRM platform and in the relevant backend system. Documents required for invoicing are either replicated in the relevant backend system or created there.

A complexity driver is definitely the "order type" differentiation between "free text" and "catalog" ordering:

- Free text ordering: It requires a higher effort during the shopping cart creation and bigger number of approvers → complexity driver.
- Catalog ordering: Predefined procedure for ordering from catalog and small number of approvers.

Fig. 12.4 shows a differentiation between simple variances vs. complex variances of a shopping cart.

An e2e transparency of the SRM P2P process can be easily provided using Process Mining from shopping cart creation until release by respective approvers that submits work item.

Process Overview and Rework—Indirect Material

The use case "SRM Overview—Process Overview" allows to analyze the true e2e process w/o any restrictions. It also serves as the starting point to explore and find optimization potentials beyond the defined use cases per country and division. This use case covers all process variations. It provides a variety of drill down information

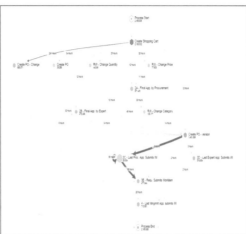

Fig. 12.4 Indirect material process complexity for shopping cart creation

to substantiate or contradict new hypotheses. The amount of Process variants is a BVI, which is helpful for displaying the amount of inefficient cases of the SRM process to be reduced to a minimum with low effort. It represents the combination of sequences between the amount of standard activities and rework activities. The number of reworks per PO item is counted based on rework reasons, e.g., change payment terms, change category, change price, delete PO item, change quantity, change currency, and revoke PO item. These reworks can be treated as potential inefficiencies of the SRM process and would therefore require an optimization. Some event types shall only be counted as rework if they occur more than once for a given process instance.

Shopping carts with number of rework steps lead to the amount of approval steps that were passed twice or more. A reduction in this number of approvers means a significant increase in operational efficiency.

Here are some example questions that shall be answered using SRM Process Mining:

- What are the reasons for rework?
- What is the rework rate by organizational unit and what is the affected value?
- What is the median subprocess throughput time by orders created as free text and for orders created as catalog?
- What is the median subprocess throughput time by vendor type?
- How many approvals are needed by POs?
- What is the number of approvals by approval step?
- What is the rework rate by procurement commodity code?
- What is the amount of approvals for free text orders with a certain value?
- What is the amount of approvals for catalog orders with a certain value?

- Which order type has more rework effort?
- Which vendor type has a greater value of rework rate?
- What is the material group that has the highest rework rate?
- How much is the price cluster for rework affected cases?
- What is the amount of rework per rework cause?

Amount of Approvals per Shopping Cart—Indirect Material
The number of approvals per shopping cart and PO item shown in Fig. 12.5 is a complexity driver for the efficiency of the SRM release process. A distribution of the number of the different types of approvals can be considered as low hanging fruits to get rid of. Examples of approver roles are procurement, shopping cart expert, and management.

A large amount of approvals can be reduced by improved information about rules of authority. Reducing effort in approvals is possible through a restructuring of the approval rules. For that, a significant percentage of the amount of the shopping cart items needing 4+ approvals can be eliminated, which can be directly linked to a permanent decrease of approvals that could be skipped.

Here are some example questions that can be answered using the approval use case:

- What is the number of approvals per approval step?
- What are the median approvals by division?
- What are the numbers of approvals by PO items?
- What are the numbers of approvals by shopping cart items?
- How many activities and cases exist per approval activity?

Throughput Times—Indirect Material
The "Throughput Times" use case focuses on lead times between key milestones as well as idle periods within workflows. The overall target is to streamline and thereby shorten the average processing time. The Throughput Time requires a defined timestamp for each process step. Approval steps always have two timestamps: when the user receives the item for approval (step start) and when the user finishes that approval step (step end).

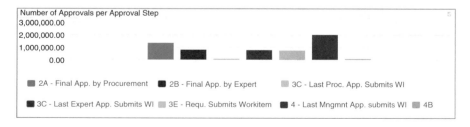

Fig. 12.5 Number of approvals per approver step

All combinations of start and end events can be usually analyzed in the Process Mining process explorer, e.g., from "shopping cart sent into approval date" until "last procurement release." For usability reasons, some combinations have to be predefined.

Here are some example questions that can be answered by the Throughput Times case:

- What is the median throughput time by organizational unit?
- What is the average throughput time by organizational unit?
- What is the throughput time by order type?
- What is the throughput time by vendor type?
- What is the throughput time by material group?
- How much time does it take to complete a PO item process?

Payment Terms

The initial hypothesis emerged within the payment term case is to consider three commonly accepted payment models and their individual impact on the cash out towards the suppliers. The hypothesis is based on the assumption that an organization uses those payment terms exclusively.

First payment term assumption An organization is fully exploiting the payment target of 90 days without any cash discount (Skonto). Therefore, all invoices which are paid earlier than the agreed payment target of 90 days provide the potential to save on capital costs. In case of late payment, rework and potential interests for late payment might be a consequence. Thus, the full exploitation of the payment target, but also the payment within the payment target, can have a tremendous effect on the cash situation of the business.

Second payment term assumption An organization is fully exploiting the 3% cash discount (Skonto) and the payment within 14 days. It is presumed that the discount is not always fully taken to the possible extent.

Third payment term assumption An organization is creating an optimized mixture of both payment target and cash discount. Thus, it uses the full potential of cash out towards the suppliers according to the local finance regulations and calculations (Fig. 12.6).

The Benefit Calculator creates transparency about potential savings with regards to the cash out towards suppliers and also with regards to the process complexity and the number of variants of the process shown in the upper figure. It is based on the

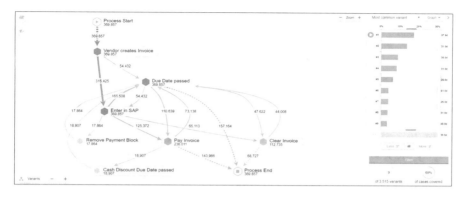

Fig. 12.6 Payment term process complexity

hypothesis that "we are not exploiting the full financial potential of cash out towards our suppliers." By exploiting the full potential of these payment terms, an organization has the ability to reduce the cash out towards its suppliers.

EVA Potential
Difference between EVA taken and EVA resulting from the configured payment term.
Formula: Theoretical EVA − EVA taken

% EVA potential
EVA potential in percentage.
Formula: $\frac{\text{EVA potential}}{\text{Theoretical EVA}}$

EDI Rate Versus Rework Rate

The digitalization of the P2P process has the potential to create value for all parties involved. As a result, the Electronic Data Interchange (EDI) introduction is key to increasing the automation rate of an organization. As the world is changing, the best way to be in line with evolution in the P2P arena is to connect with suppliers via EDI as supplier chains are becoming global with communication channels becoming even more complex.

Today, high manual effort is requested for ordering goods via e-mail or fax, which represents an outbound communication to suppliers. This process requires manual entries, and order confirmation often are being received via e-mail, phone, or fax. In addition, change requests to the order need to be communicated through the same channels, and manual changes in the system are required to get the changes done. On the inbound side, received invoices are either proceeded manually or sent via e-mail, and these invoices are in many cases scanned and verified via OCR, which leads, in many cases, to a payment delay and to additional rework caused by data mismatch.

In the future, with the help of EDI, all P2P outbound and inbound message transactions shall be managed electronically for covering message types comprising orders, order responses, order changes, dispatch advices, and invoices, which lead to increase in efficiency with speed and quality. Accurate communication of demand eliminates errors and confusion. EDI also supports standardization, agility, and scalability. The value add of EDI can easily be measured with Process Mining to display the lead time and the difference in rework rates between non-EDI and EDI transactions. A full process transparency is a prerequisite for improvement. In addition, accurate and transparent information enable compliance and security. Last but not least, such a transparency enables better supplier performance evaluation through higher data quality.

Digitizing the P2P process leads to significant time reduction potential for your own organization as well as for your external suppliers with up to 30 min per order for each party mostly by reducing the lead time of each of the major message types, e.g., order, order changes, or invoice and clarifications.

With the help of Process Mining a case has been developed to compare the impact of the introduction of EDI versus non-EDI with regards to the delivery reliability of suppliers. The case illustrates the reduction of rework activities and throughput time in dependency on the EDI rate of suppliers. It is developed based on the assumption that a "higher EDI rate results in a reduction in overall rework activities, which leads to a reduction in the throughput times" visualized in Fig. 12.7. Therefore, the higher the EDI rate the lower the rework rate.

Fig. 12.7 Correlation between EDI rate and Rework Rate and Throughput Time

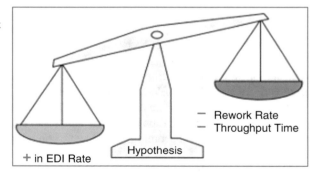

Rework Factor EDI

The factor calculated based on the overall rework activity for EDI PO items and the
total number of EDI PO items:

$$\frac{\text{Overall rework activity for EDI PO items}}{\text{Number of EDI PO items}}$$

Rework Factor Non-EDI

The factor calculated based on the overall rework activity for non-EDI PO items and
the total number of non-EDI PO items:

$$\frac{\text{Overall rework activity for non-EDI PO items}}{\text{Number of non-EDI PO items}}$$

Rework Factor Delta

The difference between the rework factor non-EDI and the rework factor EDI:

$$\text{Rework factor non-EDI} - \text{rework factor EDI}$$

Assigned Time per Activity

$$\text{Assigned time per activity} \times \text{rework factor delta} \times \text{non-EDI PO items}$$

Saving Potential in Minutes

$$\text{Assigned time per activity} \times \text{rework factor delta} \times \text{non-EDI PO items}$$

Robotic Process Automation

The robotic process automation (RPA) monitors performance through Process
Mining. The RPA goals consist of:

- Automating repetitive process activities that increases performance and reduces
  costs and errors.
- RPA captures and sorts valuable data much faster than human.
- Digitizing manual processes allow employees to focus on more complex and
  knowledge-based tasks.

RPA potentials and benefits are:

- Generate savings
- Reduction in implementation time

- Increase in operational efficiency with sustainable benefits
- Increase RPA coverage
- Better data transparency

Through the introduction of a self-service benefit calculator, it is possible to generate savings per process step or activity, e.g., "send PO update" for your organization. It is also possible to extrapolate savings through linear regression method from monetary or performance perspective in a time dimension. Key element related to P2P Process Mining is to compare two organizations with each other from different dimensions, e.g., system, country, organizational unit displayed in Fig. 12.8.

Each of the processes are being compared with each other with regards to the "manual rate," "automation rate," and "RPA rate." This enables a "direct comparison between two units within your business" and leads to the identification of potential improvement areas per P2P activity. Therefore, if in organization A the activity "create PR items" is done mainly manually, whereas the same activity within organization B is done mainly automatically, then it might be identified as an improvement measure to implement a robot for programming the RPA process. The monetary savings for the implementation can then easily be compared with the implementation costs based on real data delivered by Process Mining. The calculated business case can be taken as part of a "decision-making document" for your management in order to release the required budget for enhancement.

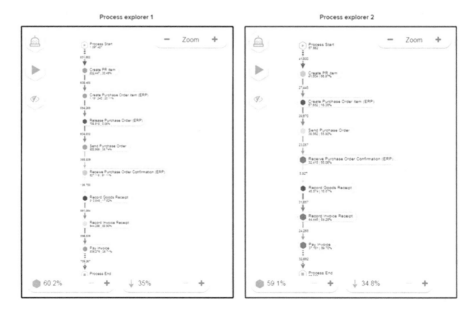

Fig. 12.8 Comparison between two P2P process explorers in two different organizations

Impact: Potential Business Benefits and Savings

The P2P Process Mining use cases can be subdivided into four major levers displayed in Fig. 12.9:

- *Increase operational efficiency*: A potential reduction of operational inefficiency can be leveraged by getting rid of inefficient activities and duplicated process steps.
- *Reduce throughput times:* Activity throughput times/lead times are to be reduced based on the process graph analysis to identify the root cause of the inefficiency by supplier, purchasing organization, transport forwarder, or material provisioning in a stock/inventory.
- *Optimize working capital:* conduct a housekeeping measure by comparing the different payment terms listed in the master data, Pos, and invoices and switch to more favorable payment terms.
- *Increase digitalization:* Introduce EDI by connecting external suppliers for the message types orders, order response, order change, dispatch advice, invoice, self-billing invoice, delivery forecast, and inventory report.

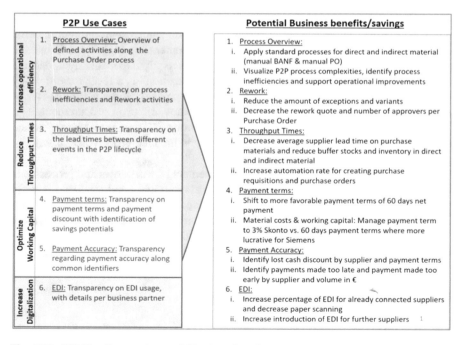

Fig. 12.9 P2P Use Cases and potential business benefits

Concrete benefits and results that have already been achieved and experiences based on P2P Process Mining will be listed in the following subchapters where measurable.

Benefits of Increasing Operational Efficiency

Using Process Mining helps in analyzing facts and findings, allowing for business consequences by identifying what the issue is that requires improvement, what would be the adequate solution and what savings potential can be reached by calculating the potential benefit. One possible option is to compare the desired process against the actual process in order to find out the root cause for deviation and performance loss. Most of the use cases for direct and indirect material of the P2P standard processes lead to areas with potential improvements when it comes to a reduction in the amount of exceptions and when the rework rate and number of approvers can be reduced to a minimum.

In one case "POs created on or after the date the invoice was issued" is beneficial to identify potential fraudulent behaviors in the P2P process. This can be easily visualized with a Process Explorer tree by selecting the dates where invoice was issued before the PO header was created.

In a further case, the "number of approvers of direct material ordering has been reduced by half" for the activities "order confirmation reminder," "PO quantity change," "PO Price change," and "PR change," which leads to higher operational efficiency. "Small value PO line items and low value changes" in indirect material ordering have in some cases same approval requirements as high value items and changes. In addition, a significant amount of effort is required to manually update the approvers for a given PO. "Increased process automation" is a lever for efficiency improvement and savings within the procurement process. Some activities can be automized in accounting, e.g., "payment block removed," or in procurement, e.g., "enter goods receipt." On the other hand implementing a "Three Way Match," which means the PO matches with the Goods Receipt and the received invoice, is a major complexity reduction driver.

For payment accuracy, a review and improvement of workflows and processes is key. Beyond that a timely goods receipt process is important and establishing daily reporting will help decrease the resolution time of payments. In other words, you should "pay your external suppliers on time according to the agreed upon payment terms" because paying them too late will lead to dissatisfaction and paying them to early will lead to a reduction of cash flow for your company or organization.

Benefits of Reducing Throughput Times

The throughput times between the "creation of a PO and recording goods" is usually measured based on the delivery date of the goods. This time span can be easily jeopardized through a rework activity, e.g., "change quantity," which normally results in an extension of the throughput time as the provider will have to delivery more goods than originally planned. Therefore, the rework rate and the throughput times are interlinked with each other. There exists a correlation between the throughput time and the rework rate, at least from PR process until goods receipt.

In another case the finding "early deliveries lead to an increase in storage costs" was identified for several suppliers in a business unit, which led to a higher rework rate and to an earlier payment to the suppliers. This case was then escalated to management and the suppliers where informed to stick to the planned delivery dates of the goods, which leads to receiving the goods according to the right throughput time.

Benefits of Optimizing Working Capital

Savings can be generated by "shifting all payments" to either lower discount to 3% within 14 days or to a negotiated payment term with 60 days or more w/o discount. For that, it is important to identify and prioritize suppliers suitable for renegotiation and to analyze suppliers with multiple different payment term equivalents and to harmonize used payment terms. Finally, it is recommended to promote supply chain finance know how within the procurement organization. This case is considered as the most efficient case as it is directly addressed to the CFO of the organization for exploiting the cash out optimization towards external suppliers from a monetary perspective.

Benefits of Increasing Digitalization

A facilitation of the full EDI usage for suppliers that already use EDI partly is considered a low hanging fruit in order to "increase the EDI rate" in percentage with external customers and by that to close the gap in the EDI rate. Furthermore, it is useful to identify the suppliers with highest impact from order volume perspective and from amount of transactions per year to prioritize those suppliers to be connected next within your company or organization. Nonetheless an EDI connection can only be established if your backend systems are ready for EDI and fully support all required message types, e.g., orders, invoices from technical and legal perspective per supplier country. Do not forget to sign an EDI agreement with the respective suppliers first before setting up the EDI connection.

A benefit calculation model can be set up for calculating process efficiency through the introduction of EDI and considering the labor costs, process efficiency in minutes/hours per message type, and region of your organization. It is proven that introducing EDI might save up to 1 hour per PO for both your own organization and the supplier organization.

Technological Concepts of Process Mining P2P

There are some technology concepts of Process Mining which need to be considered:

- *Case*: The central entity of Process Mining. Every chart, table, and process diagram is related to a case. The process flow that can be visualized must always be connected to a specific case. It can be, e.g., a PO item, a sales order, an invoice, etc.
- *Activity*: An event that occurs in a case which has a defined timestamp that can be tracked in the source system. Also defined as a process step. Examples: creation of a shopping cart, release of a PO, payment of the invoice, etc.
- *Connection*: Sequence of two activities.
- *Throughput Time*: Time spent between two activities.
- *Rework activity*: An unnecessary/repetitive step in the process.
- *Level*: Indicates if the activity is related to the PO Header (H) or the PO Item (I).

Technological Prerequisites

In order to implement P2P Process Mining within your organization, some prerequisites are to be fulfilled prior to taking advantage of this wonderful technology.

Firstly, it is recommended to have one central data source where P2P data are being collected in real time no matter what type of relational database technology is being used. A strong architecture landscape is key to a successful Process Mining implementation. Also, a data provisioning system coped with a strong service management organization that comprehends the positive impact on having service level agreements to deliver the required data to the internal customers or data consumers is of utmost importance. Therefore multiple data sources shall be normalized in a way that you are always comparing "the like with the like," no matter whether you are referring to internal data sources or to external sources that have been integrated via ETL tools, which means Extract, Transfer, and Load.

Secondly, offering harmonized data models for the different focus areas of P2P Process Mining with a centralized authority and workflow concept will definitely strengthen the data quality of your Process Mining environment in the backend development system. Implementing a library or P2P development framework will automatically lead to a harmonized "data as a service" and to a validated aggregation of calculation views, as purchasing and accounting metrics need to be displayed in different levels of aggregations per supplier, material, commodity, plant, or business segment.

Thirdly, multiple Process Mining frontend reports based on "single source of truth" support the visualization of the "to-be-defined" P2P focus areas and use cases.

All three components can either be hosted on premise or in a cloud solution. The current trend shows a move towards cloud solutions. The pros and cons of these alternatives shall be evaluated from monetary, IT security, and performance perspectives and always aligned with internal customers and stakeholders.

P2P Data Model Configuration Template

Implementing Process Mining in practice requires the expertise from data scientists to set up a project schema in which data are being structured and persisted in tables and calculation views. For that, a deep knowledge in the structure of the source system and tables for the P2P process is required for enabling the retrieval of digital traces through Process Mining. The following types of information need to be gathered prior to starting with the data source modeling:

- Scope analysis, including the legal entity structure of the organization
- Plant structure and depth structure of the organization
- Vendor account group structure, vendor classes, and differentiation between internal and external vendors
- Purchasing document type and purchasing organization structure
- Company code structure
- Classification of direct versus indirect material
- PO channels, e.g., PRs, PO direct
- Plant classification, e.g., manufacturing, sales, service plants

Additionally, process customization is required. Fig. 12.10 shows a process customization template for an SAP system.

Project Use Cases and Activities	
Technical contact:	<your name>
Entity for Activities	<your system>

No	Activity	Source	Table name	Table field (Timestamp)	Value -Optional	Example Table	Example Table field	Example Value
	Customizing SAP Messages							
	What are the relevant customizing settings in your system for Purchase Order Messages?							
1	Ord.confirm.reminder	SAP MM				NAST	NACHART (Message Type)	YHRE for Oder confirmation Reminder
2	Send PO	SAP MM				NAST	NACHART (Message Type)	YHRE for Oder confirmation Reminder
3	Send Warning	SAP MM				NAST	NACHART (Message Type)	YHRE for Oder confirmation Reminder
4	Other relevant messages?	SAP MM				NAST	NACHART (Message Type)	??
	Customizing SAP PO Approvals							
	What transitions of values mean the approval of a PO and PR in your system?							
5	Approve PO	SAP MM				EKKO	FRGKZ	Value (", X") to (2, A)
6	Approve PR	SAP MM				EBAN	FRGKZ	Value (", X") to (2, A)
7	Remove PO Approval	SAP MM				EKKO	FRGKZ	Value (", X") to (2, A)
8	Remove PR approval	SAP MM				EBAN	FRGKZ	Value (", X") to (2, A)
	Purchase Requisitions							
	How does the Purchase Requisition release take place? Is there any external workflow system or is it done in SAP? If it is done							
9	Approve PR	SAP MM				EBAN	FRGKZ	Value (", X") to (2, A)
	Purchase Orders							
	How does the Purchase Order release take place? Is there any external workflow system? If it is done externally, how is it							
10	Approve PO	SAP MM				EKKO	FRGKZ	Value (", X") to (2, A)
	Vendor type							
	How can we determine if a vendor is internal or external? Can we use the accounting group (LFA1-KTOKK) as mapping field?							
11	Vendor types	SAP MM				LFA1	KTOKK	
	Direct/indirect material and division mapping							
	How can we make the separation between direct and indirect materials?							
12	Direct/indirect material	SAP MM				EKKN	PRCTR	
13	Direct/indirect material	SAP MM				EKKO	EKORG	
	Payments of invoices							
	Which posting keys (BSEG-BSCHL) and/or document types (BKPF-BSART) should be considered as invoices for this system?							
14	Posting keys	SAP FI				BSEG	BSCHL	
15	Document types	SAP MM				BKPF	BSART	
16	Profit center	SAP FI				BSEG	BSCHL	
17	purchasing group	SAP MM				BKPF	BSART	
	Customizing Oder Confirmations							
	Do you record Order Confirmations and what are the relevant fields?							
18	Order confirmation received	SAP MM				EKES	EBTYP	AB

Fig. 12.10 Template for technical customization of the backend system

Lessons Learned and Outlook

Process Mining is a key technology of the twenty-first century, which can easily be introduced into a company and organization as it is no longer complex to retrieve digital traces of the P2P process from computer systems that are using a standard database. All that you need is the awareness about how to structure your own data with the help of data scientists that create data models and aggregated procedures of the company's master data into one central data lake and create a model to retrieve the required digital traces of the processes to visualize them in Process Mining.

From my experience, after a while the management no longer questions the correctness of the data and processes as it all becomes transparent to the purchasing heads and to the financial department down to the PO or invoice item level. The positive aspect is that this topic then becomes a pull process with higher demand over time. Nonetheless a big challenge is to focus on the right analysis and to generate maximum savings and increase in operational efficiency for the organization. The days are gone when PowerPoints were created and data are prepared for the sake of generating slides for creating decision-making documents. Through Process Mining, managers themselves work on getting the right facts on their own processes and generate the right measures. One negative aspect is the fact that "less is more," which means it is worth focusing on less improvement areas, but to implement them in a more sustainable way. Another experience from my side is the fact that not all decision-makers are monetarizing their own process improvement measures in an adequate manner.

The next decade will show a major increase in the importance of Process Mining for most businesses. A potential hypothesis is "forget about Six Sigma, now it is the age of Process Mining."

athenahealth: Process Mining for Service Integrity in Healthcare

Corey Balint, Zach Taylor, and Emily James

Abstract

athenahealth's Technology-Enabled Services—Service Outcomes team is responsible for the optimization and scaling of healthcare administration transactions—one in which we complete millions of transactions each day on our customers' behalf. athenahealth was looking to create technology tooling to gain a better understanding of total process workflows across these service lines—both legacy and new services included. With the legacy service lines, the processes were more set in place (well-known happy paths, built out home-grown tooling, known pain points and exceptions, etc.) and with the newer service, developments were shifting frequently. From all of this, athenahealth turned to Process Mining as the tool to gain clean process insights, to help improve our customer's experience, and bring more value to their practice.

Challenge

The Process Mining journey started as the leadership on the service lines were looking to go from an "understanding of their processes" to full transparency on their workflows. Team insights were created from data query tools—primarily direct SQL queries. As programs scaled, this was unsustainable from a talent, infrastructure, and consistency perspective. Leadership wanted to look at every part of the process and then break down each definable subsystem—needing to review dozens of primary workflows and hundreds of stages and steps. Leadership also strived to get a better view of bottlenecks in the system; although work had already been done to theorize where bottlenecks were and estimate their impact, they required a tool to

C. Balint (✉) · Z. Taylor · E. James
athenahealth, Process Mining Insights, Watertown, NY, USA
e-mail: cbalint@athenahealth.com

© Springer Nature Switzerland AG 2020
L. Reinkemeyer (ed.), *Process Mining in Action*,
https://doi.org/10.1007/978-3-030-40172-6_13

highlight all of them in one swoop. This is where Process Mining became a major driver.

While Process Mining was a clear tool to explore, there were a few key concerns across the organization:

- The described services are all at different stages in their life cycle—some are refined products and others are still an emerging service. Leadership needed a tool that would be flexible and could look at a product at any stage.
- On top of that, athenaNet's data model is not easily ported into a software, unlike other organizations with SAP-based systems.
- As athenahealth has a vast amount of data, there are also varied BI tools already in place—there needed to be a clear focus on process and not replicating preexisting insight tools.
- The last major concern was that the team needed to work on scaling and operationalizing as a small center of excellence (three-person team). As a part of a 5000+ person company, the team needed to not only become experts on the software and processes, but also build the knowledge on using Process Mining for process improvement.

Why Process Mining?

athenahealth first became interested in Process Mining to create a tool that would better enable root cause analysis and optimization of processes. While there were subject matter experts with metric-based knowledge about services' issues, there was no tool to quantitatively, qualitatively, and visually display how process changes would affect system metrics. Naturally there were benchmarks and KPIs to help steer decisions, but the organization lacked the "here are the detailed workflows" view with measurements around their process health. Leadership was also intrigued by the idea of viewing downstream impacts when certain task attributes were selected within the Process Mining system. As all of this information would help us swiftly bring improved value to our customers.

The lack of insights here led to diminishing returns or difficulty in tracking returns on the improvement opportunities. The Process Mining Insights (PMI) team was then established to function as an enablement team, helping operations and product teams refine their service by minimizing process variation through Process Mining tools.

Use Cases

athenahealth has rolled out two implementations across the company's service lines, with vetting of multiple services lines being conducted for future implementations.

Use Case #1

The service line the team chose for the first Process Mining implementation focused on the interactions between healthcare providers, payers, and internal operations teams to ensure patients are approved to receive the necessary treatment in a timely manner. This is a service at athenahealth that is in the process of scaling and was the ideal candidate for our first implementation for several reasons.

- First, this service needed the ability to identify and remedy bottlenecks prior to their ramp up, a key Process Mining offering.
- Second, the executive sponsor and Process Mining enthusiast had this service line in their portfolio, providing an opportunity to get Process Mining in the door.
- Lastly, the PMI team was already process experts in the service line and was dedicated to taking on future implementations and learning Process Mining techniques. This service needed Process Mining to influence decisions on both the product and operations side before broad market launch.

Use Case #2

The second Process Mining implementation focused on setting up customers to transmit claims electronically with payers, a more established service than use case #1 with more complex workflows and data structure. Unlike the first implementation where the team chose the process with the strongest executive support and where the data structure most resembled an event log, the team took a more methodical and vetted approach with the second implementation. The team evaluated, graded, and ranked different service lines across athenahealth based on different criteria such as resource availability of both process and data SMEs, data sources, RoI and a few others to ultimately decide where the biggest impact could be made. This shift to having service lines make a case to the Insights team about why they needed Process Mining tools had a significant impact on user adoption and business impact as described in detail later. In addition to the difference in intake method of the second service, the scale of transactions drastically increased to millions from the hundred thousand in use case #1, adding another level of complexity. However, as the service had been around for many years, a tremendous amount of work was already completed on mapping "good" and "bad" process flows. They just needed the added level process transparency that could only be achieved through Process Mining tools.

Implementation of Use Case #1

For our first implementation, the PMI team, made up of operations engineers, were experts on process and data but novices on Process Mining. Because of this, they followed our vendor's standard implementation procedure tightly.

- *Scoping*—putting together an event log and making sure all necessary data to calculate relevant KPIs was included.
- *Building*—uploading our V1 event log and determining which pain points should be built into which sheet (accuracy, timeliness, or efficiency).
- *Training and Go Live*—onsite training of the team and various stakeholders on how to use Process Mining techniques, making any last updates to the MVP, and going live once signed off on the build.
- *Handoff*—preparing the Insights team to take on any future build requests and begin expanding usage to end users.

Following this protocol, the vendor implementation team pushed heavily on getting the system up and running, making sure the data was reliable, and ensuring that the Insights team was equipped with the know-how on implementing future processes. The team, however, did not focus on training end users on the business teams. Instead, the team had positioned themselves as the end users throughout the implementation rather than the operational day-to-day team. Once "live," the implementation didn't have a clear "what's next"; they had over 60,000 variants out of 160,000 total cases, many different root causes at play, and only a few users knew how to apply Process Mining techniques.

As the team realized this wouldn't scale, a decision was made to take a step back and create an operational model, one that would be robust enough to apply to future implementations (for details on full operational model refer to the Key Takeaways section). Custom classroom trainings were developed to fully engage the true end users and immerse them in the world of Process Mining. Root Cause Analysis (RCA) workshops were also conducted to identify areas of inefficiency within processes for the end users to either put in place an improvement or scope further.

Example of RCA Findings for Use Case #1
An example of an inefficiency that was discovered using Process Mining was called "Administrative Swirl" as shown in Fig. 13.1. When looking at tasks that were sent to the client two or more times during the life cycle, a pattern was discovered where tasks go back and forth from the customer to athenahealth, back to the customer, and back to athenahealth, with no progress being made over many wasted business days. Through additional research, the team determined customers did not know how to update certain information, and the athenahealth user, through compliance-driven workflow restrictions, were not enabled to update the information on their behalf. When presented to the business team in the RCA Workshop, they brainstormed solutions on the spot and decided to create documentation for athenahealth users to provide the customer on the first instance of them asking for assistance. This workflow update eliminates unnecessary customer actions by enabling them to make any required updates or edits on the first attempt. This reduces their administrative burden and allows athenahealth to truly reduce the customer's workload. In addition to reducing customer workload, this also allows users to complete these tasks faster, expediting the care that patients need.

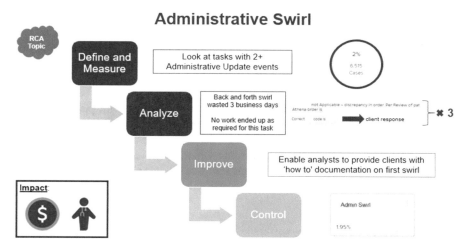

Fig. 13.1 Administrative Swirl—"DMAIC" snapshot

Implementation of Use Case #2

For the second implementation, the team had developed more expertise in Process Mining, but had little knowledge on the new process or its data. All their learning experiences from Implementation #1 were used to create a "Path to Success" framework that details the whole engagement process from start to finish (Fig. 13.2).

Using the six distinct stages of the Path to Success as an e2e implementation timeline, the team successfully implemented their second service line.

Stage 1: Exploratory Intake
Because athenahealth has many processes to which Process Mining could be applied, the first step evaluates opportunities for fit and readiness. First, the team looks at business problem(s) to be solved by the stakeholder team (with associated metrics). Second, as event log data and state models are not the same between processes, they look at the work required to collect the necessary information. The team then evaluates requirements needed for a successful implementation and scores potential investments.

Fig. 13.2 athenahealth's "Path to Success" model for Process Mining

Stage 2: Kickoff Planning
Once the decision has been made to move forward with the highest scoring service line, the PMI and stakeholder teams align on timeliness and expectations for each phase in the path to success framework.

Stage 3: Preimplementation
This stage is the heaviest lift of the whole process; it begins with putting the end users through the athenahealth custom training to get them familiar with Process Mining terminology and data requirements. Once training is complete, events and KPIs are defined and finalized to determine which events are needed to have full transparency into the process and which information is necessary to calculate all desired KPIs.

Stage 4: Implementation
With the event log now in a workable state, the team uploads test data into the software to validate formulas of KPIs, adding any missing fields to the event log as deemed necessary. Different build options are presented by the PMI team to the stakeholder team for different ways to display charts, graphs, and KPIs. Once the stakeholder team signs off, the PMI team builds out dashboards in the software.

Stage 5: MVP
Once KPIs and components are fully built and validated, full data loads are completed and followed by a thorough round of user acceptance testing with the stakeholder team. The stakeholder team and PMI team start to formalize a project backlog here with the larger business initiatives they are to be tied to. This then starts to create a "here's how we intend on using Process Mining" sheet for any future end users to refer to when starting in the software.

Stage 6: "Go Live" to Value
While this final stage is always ongoing, the team goes through a couple RCA workshops with stakeholders. These workshops are used to teach end users how to conduct root cause analyses on their own, highlight some immediate potential quick wins, and begin populating the stakeholder team's project backlog for future improvement projects.

Using this model, the team was able to reduce implementation time from use case #1 to use case #2 by 65 days and are looking to continue decreasing that time with future implementations. For each learning experience the team had encountered in the implementation, a tweak was made to the model as to not make that mistake again.

Technology

For the first two implementations, the Insights team used a variety of different technologies to successfully compile and automate event logs. As athenahealth has a lot of data, creating an event log, sourced from different locations, took a fair amount of cross team collaboration facilitated by the PMI team.

For use case #1, the data was fortunately structured in a way that closely resembled an event log—minimizing the effort needed to consolidate data. However, to ensure that all data vital to capturing customer-centric information and KPIs, the Insights team worked closely with the data SME when writing and updating the event log query. Currently this process is still manual; running the query, updating csv files, and uploading the csv files into the Process Mining software. The Insights team has plans in the coming year to automate the event log update and upload process.

For use case #2, a longstanding service line, most necessary data is stored in a data lake in which event log and attribute tables were created and automatically updated when the data lake updates. However, the team encountered an issue where certain tables that were necessary for important process data lived outside of the data lake. They then needed to work with the appropriate parties to get the tables into the data lake and then into the event log and attribute queries. With this slight setback, the team added data sources to the exploratory intake phase to proactively flag potential issues of data extraction. Once all the required data was in the data lake, with the event log and attribute tables finalized, the team set up a direct connection from the data lake to the Process Mining software for an automatic daily refresh in the software—the end state goal of the team moving forward for future processes.

Impact

Service Integrity metrics are based around three core measurement groupings:

- *Customer Outcomes*—Increase the positive outcomes for our clients (e.g., claims are processed faster).
- *Customer Work*—Reduce the administrative burden on our provider offices (e.g., reducing document touches).
- *Efficiency*—Reduce the operational costs of athenahealth work (e.g., reduce task touches or increase automation).

To give you an idea of what these all mean, when reading "efficiency"—you can think about it simply—it is the number of touches required to complete a task. These could be reduced by a simple workflow alteration or re-teaching, or something more complex where product teams our engaged to engineer changes to the system. Customer outcomes are the KPIs for the customer's work. Think of this as processing claims faster or preventing denials. Lastly there is customer work—which is exactly what it sounds like. athenahealth wants to reduce the number of times our customers need to get involved in resolving issues. athenahealth's mission is to help caregivers focus on providing the best care possible to their patients.

While it is still early in the Process Mining journey for athenahealth, both implementations have already seen value come from Process Mining's enablement. Use case #1, which has about 6 months on use case #2 timelines, has already seen ten different projects come out of the build. While some were smaller wins (prepping for industry meetings, release planning, and charter building), about half of the projects turned into significant projects or tools.

The largest internal win thus far has been the creation of a "Process Health Dashboard" to be used in weekly updates with executive leadership. With this, our executive leadership team can log in and check the service line's system health at any point in time. If they have questions on certain attributes of the system, they can filter down into those to see if there are ongoing problems. This immediately helped identify areas to drive down athenahealth cost and increase customer outcomes scores.

The largest win for our customers came from identifying a delay in task routing. The service team was able to use a process dashboard that identified one turnaround time metric whose average was spiking. From that they were able to identify what was wrong, put a new training in place, and seek out additional engagement from support teams. These fixes were put in place before any customer felt these delays.

Another win has been creating a tool to help enable automation. The tool flags certain parameters to a team that decides "is there enough good data here to build an automation rule." The team checks this on a biweekly basis to then create those new items—thus removing work off their plates each sprint. Again, another win for athenahealth cost here, but it also reduces the potential for customer work, as the larger task pool of workable items decreases.

Use case #2, only a month into their implementation, has already started operationalizing the tooling that has been built for them. Within 2 weeks of "go live" they identified an ongoing, but previously unidentified, workflow issue for a large client and were able to remedy it within another week—making huge marks on both customer outcome and customer work.

Lessons Learned

While the Process Mining Insights team has only existed for a little over a year, many lessons have been learned in that short time. The foundational lesson was that having an operational model, specific to the Process Mining venture, that spells out executing and iterating on improvement projects is absolutely necessary. The team built out guidelines to start the implementations, but as they have come through two now, the model has been refined significantly for post implementation.

RCA Workshops
The stakeholder team and PMI team discuss potential interest areas for a "deep dive" into their process models. The PMI team then does a detailed analysis around those

areas, showcasing bottlenecks and significant variations from the norm. The PMI team also shows how this information is found in the software, so the stakeholder team can replicate RCA workshops on their own.

Project Backlog
Creation of a running list of areas that the stakeholder team should review once the build is complete. These items can be spun into lists of improvement opportunities for the team to focus on. This is populated by RCA workshops, planning discussions prior to the build, and items that come up once the stakeholders are using the software.

Pulse Checking
- *Enhancement requests*—a tool that's available for all end users to request functionality (new or revised) to their current implementation
- *Monthly management meetings*—system for stakeholder team leadership to check in on initiative statuses, make sure Process Mining projects are resourced appropriately, and create transparency on adoption
- *Retrospectives*—periodic reviews to discuss current functionality and usage to see how the experience can be improved

Community Building
- *Working Group*—weekly check-ins with end users to prioritize and assign items on their project backlog, read out on findings, and discuss workflow improvements
- *User Channel*—internal athenahealth community where viewers and analysts can ask questions to be answered by the PMI team or any other end users

Outside of the operational model, the PMI team also had a focus on how to properly scale and make future implementations more efficient. These improvements can be categorized between macro- and microlevel alterations (discussed next).

Macro

Team Engagement
In the first implementation, the PMI team had been positioned as the end users; ultimately this hampered end user adoption as they weren't quick to pick up the tool. For the second implementation, the stakeholder team was positioned as the end user, and the PMI took them through the entire build as a joint team.

Build Efficiency
PMI generated templates to use for all future processes—creating simple building blocks to work off, including:

- Get help and ask questions sheets
- Filter elements that can be transported from analysis to analysis
- Tools (benchmarking and swirl analysis) that can be applied to any process

PMI has now started to use the concept of "storyboards" before creating the actual build, ensuring alignment off the bat while greatly reducing the chance for rework.

Micro

Event or "Node" Identification
Choosing the proper events to focus on is challenging; it is a continual conversation throughout scoping and build. The list of events to focus on will change, but if it is discussed, it shouldn't become problematic.

Project Area and Focus
With a working knowledge of the software (post first implementation), PMI was able to have deeper discussions on "what are we trying to find a solution for," thus creating a better tool for the end users. While building to general problems and KPIs is a good start, diving deeper into "how will you as a user interact with this," is a question to always be asking.

Key Takeaways

athenahealth's early-stage success can be broken down into four Process Mining transformation takeaways.

Creating Hypotheses
First, create a hypothesis for issues you want to explore. Diving into the Process Mining tool hoping to stumble upon issues will absolutely be successful. However, having a set issue in mind and then exploring it will save you time and a likely headache as you decide whether the "issue" is material to your business's primary concerns.

Building Strong Partnerships
Keep strong partnerships with all the groups you work with. Having the data team, product team, and operations team aligned on intent and goals will keep the program moving forward. Be the group that enables open conversation, even if you do it individually to start. These partnerships allow you to create the one list of wants and needs for all to align on.

Defined Improvement Structure
Define your improvement structure. Everyone sees new projects and tries to take them on immediately. If you provide them structure around how to approach findings and then action them, you'll be able to control the scope creep better, and you will end up with a tighter RoI story when the time comes to demonstrate value.

Involvement in the Action
Be involved in the action with stakeholders. While you might not be making the changes yourself, being on the ground with the teams will keep that partnership strong and keep teams focused on improvements. Starting with these, your organization will be setting itself up with a structured format primed for consistent and meaningful improvements.

Outlook

Process Mining is a catalyst for change in an organization—systems can use any Process Mining tool to create a starting point on their current process health and start building their backlog for change. While there are some baseline requirements (a good data system and understanding of your data), once you have the data you need, you can start expanding your toolsets and skills. Outside of fixing bottlenecks, one can start building out active alerting systems, detailed dashboarding tools, or start using robotic process automation to remove repetitive work. All of these enable massive potential for increased value brought to clients and downstream savings for athenahealth.

While Process Mining is better understood with standardized process examples like order to cash and procure to pay, it is just as effective in more complex processes in a variety of industries. No matter the process, having full transparency into your workflows is what every process owner should strive for.

Links and Background Information

Who is athenahealth and the PMI team?
athenahealth® cloud-based services partners with ambulatory healthcare organizations to drive clinical and financial results. Our vision is to create a thriving ecosystem that delivers accessible, high-quality, and sustainable healthcare for all. athenahealth pursues this through medical record, revenue cycle, patient engagement, and care coordination service offerings. Expert teams build modern technology on an open, connected ecosystem, yielding insights that make a difference for our customers and their patients. These services include medical billing and practice management (athenaCollector®), electronic health records and clinical documentation (athenaClinicals®), and patient engagement tooling (athenaCommunicator®). This cloud-based/network model allows athenahealth to aggregate work, identify best practices, and optimize outcomes for more than 100,000 physicians and other

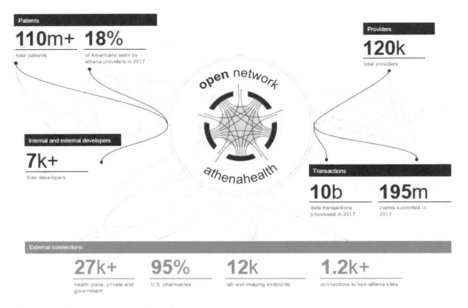

Fig. 13.3 Snapshot of athenahealth's network details

healthcare providers. Because all customers are on a single data model and infrastructure, athenahealth has unprecedented insights and knowledge into the userbase. Ultimately, athenahealth can use this network to work towards achieving one of the biggest goals of the healthcare industry: interoperability.

To ground you in the scale of the potential work—from a claim perspective, about 195 million claims are processed through the athenaCollector engine with close to 10 billion transactions in just 1 year. athenahealth is connected to 95% of all US-based pharmacies and about 27,000 different health plans (Fig. 13.3).

The Process Mining Insights (PMI) team currently sits within the Technology Enabled Services (TES) division at athenahealth. PMI leverages technology to empower teams to work more effectively and efficiently so that they can focus their energies on high value tasks for athenahealth customers. Specifically, PMI is in the "service optimization" branch of TES—whose overall goal is to improve the scale and execution of our services to deliver optimal outcomes to our customers and make athenahealth more efficient. Service optimization is composed of:

- Configuration services—which is setting up customer connections to payers when they are onboarding onto athenahealth services
- Revenue cycle management—specifically focused on the claims life cycle
- Clinical information—any core service that contains clinical information

EDP Comercial: Sales and Service Digitization

14

Ricardo Henriques

Abstract

Energias de Portugal (EDP) is aiming to become a digital utility provider. Process Mining plays a pivotal role in the digital transformation journey and helps to transform the sales to debt cycle including onboarding, billing, debt management, and customer care. It provides insights into real-world activities and customer behaviors that help to reshape the way to do business. Customer experience visualization and cross-silo transparency allows new ways to analyze actual processes and provides a foundation to boost business efficiency.

Challenge

EDP is a Portuguese electric utilities company, headquartered in Lisbon and founded in 1976. The EDP Group's activities are centered on the generation and distribution of electric power, as well as the information technologies areas. EDP Comercial, the Portuguese electricity and gas seller from group EDP, that is committed to creating a more sustainable world, is using Process Mining to kickstart a digital transformation. Enterprise-wide initiative goals include business agility across departments, achieving efficient front-office and back-office operations, and delivering faster, more reliable customer service.

With 74% of its generation capacity coming from renewable resources, EDP offers its customers energy solutions that are digital, decentralized, and decarbonized. Using Process Mining, EDP Comercial analyzed over 5 million transactions and 1 million customer-facing operations with newfound transparency, enabling to pinpoint root causes of issues that could be solved to optimize operations.

R. Henriques (✉)
EDP Comercial, Lisbon, Portugal
e-mail: Ricardohugo.henriques@edp.com

© Springer Nature Switzerland AG 2020
L. Reinkemeyer (ed.), *Process Mining in Action*,
https://doi.org/10.1007/978-3-030-40172-6_14

109

The energy retail business management must balance the following main challenges and goals:

- Increase customer satisfaction
- Reduce cost-to-serve
- Increase sales with a strong time-to-value
- Ensure compliance and regulatory obligations
- Promote sustainable offers supporting decarbonization

The market is evolving not only to sell electricity or gas, but value-added services that increase customer loyalty are of increasing importance. It poses a particularly big challenge to manage all these goals when the traditional approach of working with process mapping and business intelligence tools generates a silo-based vision of processes, slow and difficult root cause analysis, high dependency of technical skills analyzing processes with big amounts of data. Prior to using Process Mining, front- and back-office business operations were daily stressed with new issues to solve and the capacity to provide satisfactory answers was limited.

Process Intelligence aims to use the Process Mining technology that accelerates and gives transparency about processes. Processing huge amounts of data in a quick way was an important requirement. The objective was to create a new way of working around business processes with sophisticated methods that could explain in detail how processes are actually performed. And combine it with other technologies like RPA to reach a business autonomous stage of control and act with maximum process efficiency and a high-performance time to value.

From a technical perspective, the sheer volume of data to process in the last 2 years was a particular challenge, with over 240 million events representing over 200 activities.

Use Case

EDP Comercial started using Process Mining for insights regarding all B2B and B2C processes throughout the Sales-to-Debt cycle. Business cycle operations of sales are critical to ensure smooth customer onboarding, billing, collection, and debt management. About 6.1 million customer orders per year have a direct impact on the satisfaction of around 4 million customers.

In 2017 EDP Comercial—after assessing several Process Mining tools and using a trial version—implemented a first use case for a proof of value in the B2C Invoice to Debt cycle. This first case allowed to understand customer interactions from sending out invoices till debt collection.

In 2018, the implementation project for Sales to Debt and Customer Care—both for B2C and B2B—was started, with an initial focus on the B2C operations. The consumer interaction activities included onboarding, billing, collections, debt management, and customer care. All B2B and B2C processes throughout the Sales-to-Debt cycle were interconnected, thus allowing to analyze and control process performance, identify process improvements, and increase automation. Error-handling times and operations service levels were made transparent. In addition, Process Mining helps in regulatory and compliance issues by providing new levels of objectivity and transparency in its processes.

The implementation of Process Mining at EDP Comercial had different stages of maturity:

1. Trial: very fast initial trial, only trying a very small data set for a clearly specified process
2. Proof of value: allows to understand how the technology works and subsequently apply it in a specific process to create the business case
3. Project: to expand the initiative towards regular operations

Some examples of our use case activities:

Invoice overdue
Meter reading created
Meter reading taken
Billing calculation document created
Invoice created
Unread reading calculation document
Late meter reading
Resolution of request
Request creation
Total payment
Timely meter reading
Billing inquiry
Charge generation finished
Interruption notice
Meter reading query
Charged meter reading
Partial payment
SMS reminder
Sales order registration
External billing marking
Communication of meter readings

Impact

During the implementation project the internal team that was working with Process Mining described the following types of general results and impact:

- *Analytical speed*—The insights provided with Process Mining reduced the time required to analyze critical processes issues.
- *Less effort* with operational report—Due to an easier access to reports and analytical data, significant personal effort and hours could be avoided.
- Increased capacity to *deal with business changes*—The time to value, defined as average time required for discovering a new process, implementing new controls or process improvements, could be reduced. This facilitated more capacity to support changes in business.
- *Clear vision* about what happens with processes—Process Mining allows to understand actual process flows and operational performance. This insight allows to boost Net Promoter Score (NPS) results, reduce working capital and cost-to-serve
- *Anticipate problems* with operational alarms—measuring occurrence and quantity of noncompliant events allows to identify process problems. Consuming process alerts allows to anticipate issue and support preventive actions.

Medium- and Long-Term Impacts: Accelerating the Positive Impact on Climate Change

The above results and impacts combined represent an important step for business transformation through digitization, which is accelerating the decarbonized and decentralized energy trends. Prosumer clients with decarbonized concerns are demanding new offers and experiences where Process Mining plays a key role supporting this transition, which would be slower with the traditional approach of process mapping. The Process Mining transparency provides new insights about customer behaviors and the global experience that they have, which facilitates decision-makers role with real facts about their business supporting better decisions and most experimental capabilities to remake business more adaptable to the market needs.

Customer Care Insights

Deep dives into customer care insights from Process Mining provide a strong opportunity to understand in detail what were the main root causes of certain process issues. Insights allow to identify the happy path, select specific bumpy paths, and facilitate lessons learned (Fig. 14.1).

Happy Path Average customer care service on targeted SLA.

Bumpy Path Outliers are difficult to explain and we have to learn about rework loops (not visible in the operational reporting) and bottlenecks to improve performance.

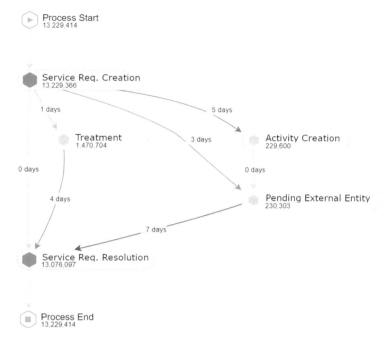

Fig. 14.1 Process sample

Lesson Learned visualizing the entire experience helps the SLA alignment between operational teams. They will not look inside their process but also understand what is happening before and after their steps (Fig. 14.2).

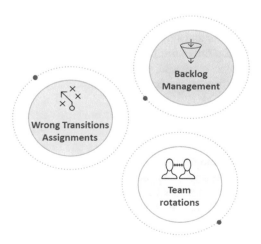

Fig. 14.2 Root cause analysis

Insights allow a new form of root cause analysis: the way of managing the operational backlog of service requests (FIFO, LIFO), the dispatching for back-office teams, including wrong transition assignments and the team rotations were identified with Process Mining techniques and helped customer care supervisors taking new actions to improve the operational performance based on facts instead of perceptions.

How Process Mining Is Contributing to Break Silos?

The interconnected vision between processes represents a new way of measuring the operational performance. When a process owner has the capacity to understand what is happening before and after his own process, a new insight appears and some key performance indicators to measure process performance could be revised with different inputs from other processes. On the other hand, the collaboration between process owners wins a new life as they have increased the capacity to control each process not only with a vision of the processes in their own responsibility, but also in close collaboration with peers in charge of prior or subsequent processes. When we link different processes in an interconnected way, we understand the link, e.g., between the sales and the contract phase, which allows our teams to reshape and rethink the KPIs that they are using. Another great example of this kind of process owner's collaboration is the joint definition of process alerts that detect potential issues based on new business rules provided by the impacted process owner instead of the real process owner. Today, these tools are used to analyze and control process performance, identify process improvements, and increase automation throughout in order to reduce error handling time and improve operations service levels.

Analyzing Within the Processes

The capability of benchmarking inside the same process using different context is especially useful to understand how the process owner could make some process improvements to increase the entire process performance. Globally we could summarize two main results: increased analytical capacity of the operational teams (analyzing and monitoring processes) and transparency regarding the executed processes.

The easy-to-use visual analysis has helped business users take actions to keep processes flowing where they may previously have diverted. This increased understanding has helped the team develop benchmarks and meaningful KPIs, ensuring a data-driven approach to understanding and improving their performance in a way that business intelligence solutions alone could not accomplish. "Our team of business users have gone from 'it's impossible' to 'how is this possible?'" Process insights has allowed to create an accelerated, efficient, and more collaborative environment with quicker, sharper decisions and to adapt to changes that wouldn't have been able to detect without Process Mining.

Technology

Our Process Intelligence platform is mainly composed of two components: the backend is a SAP HANA database where we perform Extract, Transport, Load (ETL) using SAP HANA Smart Data Access (SDA) and SAP HANA Smart Data Integrator (SDI) for connection and extraction of data from our different source systems (Fig. 14.3).

Regarding the frontend, we've a Process Mining tool as our main application, but also some e-mail alerts sent by SAP HANA XS regarding exceptions that were found during the process analysis. In those cases where we experience difficulties in accessing the data from specific source systems we use RPA bots to collect selected data from operational reporting.

Lessons Learned

About People

As our CEO said, "Digital is a people thing," so people play a key role to put Process Mining into action, which implies that it is crucial to select the right staff to deliver these kinds of projects. Concepts and technologies are typically new and differ from traditional BI tools or the traditional business process mapping approach. The following list provides some selected skills that are of importance:

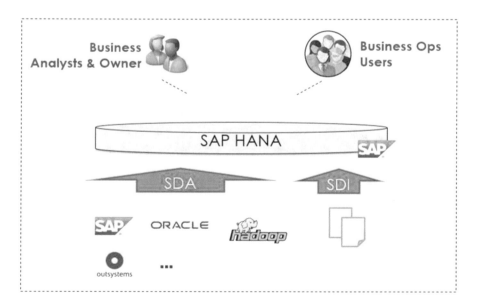

Fig. 14.3 Process Intelligence platform

- Functional knowledge about business processes combined with technical know-how regarding accessing the relevant data.
- Capacity to learn, but also to unlearn (namely, to step away from traditional ways of working with BI and process mapping).
- Agile mindset collaborating with multiskilled teams, delivering new capabilities in each Scrum sprint.

Furthermore, it is also very important to create and establish a Process Mining community with regular communication, as business users will always be on different maturity levels and continuously need to feel that they are working together towards a common purpose. The upskilling of Process Mining knowledge can be implemented in different ways, through online or on-the-job training, webcasts for specific topics, workshops, and open sessions.

About Processes

Processes with a high level of operational control through other tools tend to slow down the Process Mining adoption. Users working with established alternative tools will move rather slow compared to those users that are starting based on Process Mining insights. Accessing data is a key success factor as insights will not be valued if they do not reflect the complete flow of activities. Therefore, it is highly recommended to ensure that data collection will not be a critical issue and that appropriate access to digital traces can be granted.

How to Engage?

From our experience it is relevant to use Process Mining rather as an improvement tool instead of an auditing support instrument. Engaging operational teams to analyze, monitor, and improve their processes with a collaborative approach will work better than impose a new tool for auditing reasons.

Outlook

The Main Role of Process Mining

First, Process Mining plays a key role supporting digital transformation. Each company that is supporting the digital transformation initiatives with Process Mining insights is growing the seeds which will yield a lot of benefits in the future. Without knowing how business processes are actually happening the digital transformation journey will drive companies far away from their business reality.

Public Sector Benefits

Applying Process Mining to public sector institutions processes will allow to boost public services, remove operational waste, increase citizen satisfaction, and reduce taxes.

Process Mining Requires New Skills

The continuous growth of Process Mining has a strong relation with people upskilling, as people will have to unlearn traditional ways of thinking and learn what is Process Mining, how it works, and how different it is from other technologies.

Technological Context

From a technological perspective, Process Mining is allowing synergies with RPA or low code frameworks and will benefit from evolving AI capabilities. Process Mining will not only be used to identify new potential automations, but also to identify process exceptions that will give work to a future digital workforce (RPA) or support to identify robots that are not doing what is expected. This could mean that Process Mining will become part of a puzzle of the intelligent enterprise, where discover, analyze, improve, and automate processes will be the DNA of running business in a more autonomous and efficient way.

Organic Context

Process Mining will also play an important role to increase collaborative intelligence initiatives, breaking organic silos and increasing the collaboration between process stakeholders. Considering that companies are moving from command-control models to agile structures, Process Mining will be a critical success factor for this transition.

Is the Global Market Mature?

Even knowing that Process Mining is not a new technology, there is a long journey to go from an organic and mindset point of view. How many companies have their top decision-makers using Process Mining to support strategic decisions? How many companies have this kind of different analytical way of thinking impregnated? How many companies are calculating the real business outcome which is provided by Process Mining? The current vision of the global Process Mining market discloses that companies are shifting from a "Discovery" phase to an "Enhance" one, which means that companies are now moving from Insights to Action.

Links (Fig. 14.4)

More details about the success story of EDP Comercial:
The importance of breaking silos to boost customer experience: https://www.
    linkedin.com/pulse/breaking-silos-boost-customer-experience-ricardo-
    henriques/

Fig. 14.4 Sharing experiences about Process Intelligence @ IDC AI & BigData Leadership Forum 2019

ABB: From Mining Processes Towards Driving Processes

15

Heymen Jansen

Abstract

ABB is a global technology company in the sector of Power and Automation and is using Process Mining technology for improving its performance towards four key performance indicators: Care, Customer, Cost, and Cash. In order to get the maximum out of this, our ultimate goal is to use Process Mining not just for analytics purposes, but ultimately to use it as a technology, supporting the business in early identification of opportunities and risks, so moving from *Mining* processes towards *Driving* processes.

Starting with lead times and on-time deliveries, Process Mining has expanded at ABB towards hundreds of analytics cases: from logistics to finance to manufacturing. This innovative technology is used extensively throughout the organization and supported with an elaborate governance model, assuring continuous improvements.

Challenge

ABB people are proud to work for the company, committed to deliver results to its customers, whilst of course also facing daily challenges, which is part of the evolution of every company, whether it's a small local family business or a multinational giant. As a technology company, ABB, since its founding, has been in a continuous process of improving its product portfolio, quality, and market presence. And from the start, driving technological innovation is in the hearts, minds, and genes of every ABB employee.

So, managing change is what we're used to at ABB. There is a well-established continuous improvement (CI) culture since decades, with a clear focus on finding

H. Jansen (✉)
ABB Group VP Global Business Services, Leeuwarden, Netherlands
e-mail: Heymen.jansen@nl.abb.com

© Springer Nature Switzerland AG 2020
L. Reinkemeyer (ed.), *Process Mining in Action*,
https://doi.org/10.1007/978-3-030-40172-6_15

opportunities of improving quality, taking out waste, through the entire organization. Strict targets are being set yearly on OPQ (Opportunities for Perfecting Quality) and related to those opportunities, driving thousands of improvement projects, leading to higher quality and efficiency, in order to retain our leadership position in an ever more becoming competitive market.

But it's not all glory of course; as in many other multinational companies there are challenges as well in this journey. As one of the first companies, ABB has embraced Process Mining as one of the key enablers for fueling its continuous improvement culture, and that is what this chapter is all about. This does not happen overtime but includes piloting and local rollouts, increasing the organizational scope to some global business lines and then finally take the global approach. It takes time, but it's also a lot of fun! But most important: creating value and showing, not actually demonstrating it to the rest of the employees within the organization, which did not have the chance yet to explore the almost endless opportunities Process Mining, is needed every single day, since It's a fairly new technology and concept. And alike any new concepts, that takes time, continuous dedication and focus and above all, delivering results.

The internal and external supply chain networks in ABB contain many process variants, due to the broad product portfolio and its related manufacturing typologies like Engineer to Order, Configure to Order, Make to Order, Make to Stock, including also service process variants like field service, deport repair, and spare parts sales.

In addition to that, due to a long history of successful mergers and acquisitions, there is another side to the coin, of inheriting new process setups and (ERP) systems as shown in Fig. 15.1. This adds to process variance of course and by nature contains in the beginning many inefficiencies between the new individual partners, potentially across the entire e2e value chain.

That's not directly a problem in itself, but needs to be managed in a structured and governed way. So this needs continuous attention of all resources involved in the change, in managing the (new) business in the most lean way. But identifying and eliminating these inefficiencies in processes and systems takes considerable effort of human resources, particularly in a multidimensional systems landscape. And thus, the only option for getting better control on real-time processes within the e2e supply

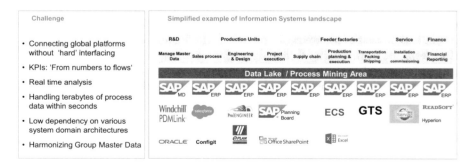

Fig. 15.1 Challenges and landscape

chain, is to create an "umbrella information system, connecting all the individual dots."

But first things first: we needed to get the connections in place for over 40 ERP systems across more than 100 countries since in Process Mining, collecting data is the first thing to do if you want to know how your company looks like in real practice. The second step is to interconnect data across countries, systems, processes, and business organizations. That implies that you also need to have a well-managed structure of master data management across company systems. And what is a good benchmark; when is the data quality good enough to start? So you might think: what's the sense of collecting huge volumes of data with the risk of having data we cannot fully trust?

The answer is simple: transparency on bad data is the only opportunity to improve it! Process Mining is not only useful to explore processes, it's also beneficial in improving data quality. So don't wait until you have the 110% perfect data in place. Learn, improve, and learn more and you will get there much faster than when complaining, talking about it, waiting for someone else to take care of it. That's not how it works best.

Use Case

Why to start? How to start? Where to start? When to start?

It of course always starts with either a problem or an opportunity, dependent on how pessimistic or optimistic you look at it. In most cases problems and opportunities are interconnected, but anyway there must be a compelling reason in considering using Process Mining as a technology to work on a specific improvement objective, whether it's in improving quality, efficiency, or cashflow. In our case, we wanted to specifically improve our Lead Times and On-Time Delivery towards our customers. Is that special, innovative, first time ever done? No its not, but it's simply our obligation to our customers.

So a simple start (you think).

Maybe not. If you want to analyze global processes and related performance for more than 40 ERP systems, which were not necessarily set up from the start in the same way and have had some considerable customizations implemented in the past due to probably very good reasons, you need to make sure that you compare apples to apples. And take into consideration that due to a wide product portfolio, there might be a lot of different apple types to benchmark.

In terms of uses cases, over the years hundreds of analytics have been developed for various reasons. Partly based on a specific need from a global business line or from a local plant. In principle both are fine, since both can bring insights which might be valuable for others within the company. It adds to creating specific process insights and benchmarks between best-in-class businesses and others to learn from those. And the good news is that there is always a unit outperforming others in a specific area and in this way, by sharing knowledge and experience on how to improve performance in a process domain, a culture of continuous improvement and interest in learning from others is established and kept alive.

Data models and analytics have been built on processes like P2P, O2C, Opportunity Management, Planning and Fulfillment, Procurement, Logistics, Accounts Payable and Receivable, Quality Management, Shop Floor Management, etc.

Use cases in the beginning typically relate to discovery of where potential issues arise:

- How are Payment Terms aligned with actual payments?
- What is slowing down the delivery process toward customer deliveries?
- Where do we see high inventories and is it related to a specific product, a market, or a country?
- Where do we have low utilization of resources in the factory?

In the next phase, this evolves into:

- How to create the optimum on cash performance?
- How to optimize the inventory planning parameters, to decrease inventories?
- How to increase the overall utilization of resources in the factory?
- What's the most optimal mix between customer service and costs?

After that, questions arise on:

- How to improve the overall lead time through the entire e2e supply chain?
- How should the manufacturing and distribution footprint look like?

These are just a few explanatory examples and might be different for various industries, but it shows that Process Mining is typically an area in which we try to understand why things are happening in a certain way and not in a different way. It's a journey in continuously searching for the optimum.

Impact

Did we have a strategy in place? Was it trial and error? Did we get directly a good return of investment? Probably a good mix of it. And it worked out pretty well, though there is always room for improvement, and we are always extremely interested to learn from others, doing things in different ways and successfully.

First of all it's pretty obvious that you always need to push hard for results and sometimes you need to have a bit of luck as well. If that happens, my advice would be: stay humble. Some people tend to call it "Our strategy works!" when gaining some results, and when things do not turn out right, they might like to change the message in "Continuous improvement takes time!" (Fig. 15.2).

And actually both might be true, but most likely it's just hard work with all kinds of ups and downs. Goethe has coined the inspiring phrase: 'Everything is hard before

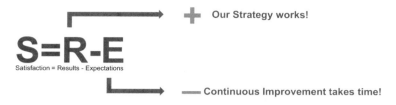

Fig. 15.2 Strategy and continuous improvement

its easy.' And that is exactly where it started at ABB. First it started with an individual believing in something new, something which might work, worthwhile trying. Then others got inspired and after a few years, we can truly talk about having a Process Mining community. Still young, still volatile, but a growing community inspired by showing transparency and opportunities in improving the business.

That pioneering approach made ABB one of the first companies using Process Mining in a real business practice, after it was given birth by the universities more than 10 years before.

Whether there was a strategy in place probably depends on the definition, but for sure there was a lot of common sense thinking in it.

- You need to know where you are and where you want to be (monitoring).
- You need to know how you can get there (process improvement).
- You need to know how to execute on it, to make sure that your moving in the right direction (Fig. 15.3).

DigitalOps Combines Three Domains

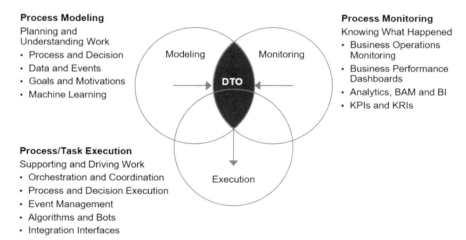

Fig. 15.3 DigitalOps and Digital Twin of an Organization

So in our case its very much about implementing the Gartner model of a Digital Twin of an Organization (DTO), which actually makes a lot of sense with regards to Process Mining technology. We all know the model of the Digital Twins, when we talk about airplanes, production lines, cities, etc. But a DTO, does that make sense? Yes, we think it does. It's an unique opportunity to reflect on what is happening real-time in the organization on many management levels and in many processes. Process Mining shows you how you manage orders, inventories, cash, etc., and so releases a wealth of information on where a certain process can be improved.

But more importantly: what you can measure, you can also simulate. So in the end you can use Process Mining tools and techniques to make the transition from Descriptive Analytics towards Prescriptive Analytics. And then it may not be too big a step towards using Process Mining for triggering planners, buyers, controllers, etc., to act on a certain change somewhere in the ordering process within the entire e2e supply chain. And in a multinational company with many internal and external trade relationships, this is instrumental for driving speed in customer deliveries and reducing manufacturing costs and inventories, due to a much better communication and collaboration process and more timely responses in this entire supply chain. Any change in the chain will be detected and via ML, AI, RPA, etc., employees will be informed much earlier to take an action on that change.

That requires of course a much higher frequency of data actualization. For standard analytics, it might be sufficient to do a weekly refresh of all data, but when using the same data for triggering required actions on changes which happened in the whole workflow, you need at least a daily data load and preferably refresh the data a couple of times per day.

Technology

Handling high volumes of data requires of course some thinking how to make the data available. As ABB started already quite early on with the journey of Process Mining, with a fairly small system, organizational and regional scope, there was sufficient time to set up the required infrastructure and support organization. Things got more difficult when we moved to an enterprise-wide deployment, with over 100 countries, more than 40 ERP systems, and hundreds of reporting units. And since building the connections to the extractors was done within quite a short time frame of less than 6 months for the majority of systems, after a few months we were already facing longer response times, which in the end are not extremely helpful when ramping up the usage of the tools globally, resulting in parallel onboarding of hundreds of users weekly. Balancing the ability to extract and visualize large amounts of data, while keeping user satisfaction high, means making choices on the relevance of data:

- Is all data really required or can it be limited to the minimal needed, based on the business requirements and/or use case?
- Can the time horizon be limited, based on the use case, i.e., for a short life cycle business you might be able to limit the time horizon, whilst for large businesses you might need a longer one?
- Is there overlap in some data models, which can be optimized by merging them?
- Is it necessary to drill down to many details (specifically on the activity level) or is it sufficient to limit the number of activities shown to the real crucial ones? (Fig. 15.4).

At the moment the data is stored on premise on two physical servers, but it is to be expected that within the next year(s), this will be migrated to a cloud-based solution, specifically to accommodate near-real-time extraction of data for the digital supply chain. New extraction techniques allow us to do so, with much less ERP system occupation as well.

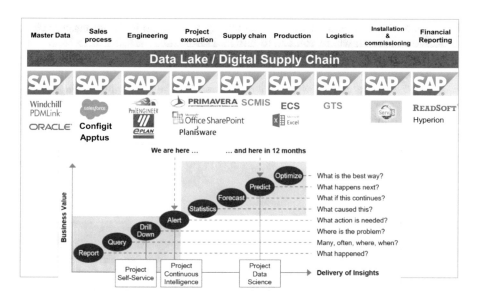

Fig. 15.4 Architecture

Lessons Learned

Experiencing a new technology means learning by doing. And in the beginning we did not experience too many best practices we could learn from. The best way for us was to think big, but to start small and act fast. Moving from some local use cases on P2P and O2C into a global deployment of covering all manufacturing sites, service centers, distribution warehouses, and sales offices, including a very wide product portfolio has of course its challenges.

One thing is key here: governance (Fig. 15.5).

In an extensive program of deploying Process Mining on a global scale, it is crucial to have a clear governance structure in place. The structure itself is not the objective, but it supports the goals, the direction you're heading in, and the common objectives to become better, together. It's a collaboration model, so the key is to make sure that all parties are involved, are aligned and, particularly, are aware of their roles and responsibilities and that those are not mixed up. In general, without going into the specifics of the technical aspects of a role/responsibility, you could divide them simply into three parts:

1. Responsibility as producer (sleeves up and make it happen)
2. Responsibility as expert (speak up when you think you can add value to the discussion due to your expertise)
3. Responsibility as team member (if something tends to go in the wrong direction, support the team where you can to mitigate risk)

Our governance model worked well and still is key in value-adding to the businesses, to create clarity on responsibilities and, by all means, to create a culture of being together in the journey of our well-established continuous improvement process, but with entire new tools, techniques, and attitude.

So, the main transformation drivers for us were and still are the following.

Fig. 15.5 Governance

Management Attention
Yes, management attention is needed. A compelling case is needed. A clear corporate direction is needed. At ABB, when the global initiative for deploying Process Mining worldwide was initiated, the corporate management chose to use Process Mining not only for analytics but also for corporate reporting, which basically had two benefits: (1) eliminating costly manual reporting efforts, by taking inputs and results directly for the core ERP systems, so basically directly at the source of where processes are being managed, and (2) connecting it to scorecard elements in order to create a natural desire in all layers of the organization to improve processes, systems, and data quality at the source.

One of the drivers has been to automate the reporting on some of our corporate KPIs, which helped a lot to make sure that all levels of the organizations were working on the same goal and using Process Mining to make it a bit easier to reach those goals.

Continues Improvement Programs
Through aligning our continuous improvement initiatives and a well-coordinated and structured CI program with Process Mining capabilities, we enabled our experts in the factories, service units, distribution centers, and sales offices to run their improvement projects faster and more effectively. If you run a traditional DMAIC cycle, it is obvious that the first steps of Define, Measure, Analyze take time when there is no full transparency of what's happening within the organization, whether it's on a local or global level. So transparency through Process Mining capabilities supports the DMAIC process enormously and specifically supports the experts to carry out their improvement projects with less effort, more focus, and better results.

Our experts are everywhere locally and globally and adding value to the company in many different ways. It's about empowering these experts.

"Performance is everybody's job!"

Creating a Future Perspective
People need to have a goal, a dream, future perspective. So we need to show a clear vision of what the future might look like and how much better things can be than they are today. Not just to stimulate or to excite people, but to show potential. Realistic potential of course. That can be done by small, but real use cases: to show that it works on a smaller scale and by enlarging that scale that the potential is there and that they have an important role in it. Maybe not having full insight in what's happening throughout the entire value chain, but to envision that they are a crucial part of that chain.

And most important: that the supply chain of the future is digital (Fig. 15.6).

In terms of lessons learned, it's important to mention that a step-by-step approach is needed. First to make sure that the transparency created by Process Mining is clear to all people involved within the organization and that there is time to "absorb the unexpected." Not everybody within the organization is necessarily very happy with transparency.

Fig. 15.6 Continuous
improvement

I. Local Continuous Improvement

- Transparency showing Improvement Potential
- Increased speed of Improvement Projects

Cultural Change: 'Performance is everybody's job'

II. Global Business Transformation

- Leveraging the Digital Transformation of the E2E Value Chain
- Improve operational activities and increase transparent collaboration
- To increase Speed, Quality, Cost and Cash
- By Lead Time / Inventory Reduction, Process Quality Improvement, Process Standardization & Automation, Footprint Rationalization

Supply Chain of the Future: 'It's Data, it's Digital'

Outlook

Process Mining is still a relatively new area in data analytics, but it is evolving fast. There is still a lot to explore, to try out, and to implement. Most likely there will be a bright future in using Process Mining in our digital supply chain. It is becoming clear that there is enormous potential for the company for further improvement and digitalization in many areas, whether it's in the domain of Customer, Cost, or Cash. Process Mining was started in 2013 as a bottom-up initiative and was transformed after a few years into a top-down program sponsored by the top management. That helped a lot in driving change throughout the organization. We have achieved good results so far, have a good buy-in of our businesses, and have already embedded the principles of Process Mining across the company.

New Process Mining technologies are opening up new opportunities. Being able in the future to work real time means that collaboration between partners within the e2e supply chain can be optimized. Our employees can focus on making the right decisions based on descriptive analytics, instead of spending a lot of time figuring out what changes happened over the last day. It will reduce communication effort, speed up the actions to be taken to keep the process running, and will probably even increase employee satisfaction. So yes, the supply chain of the future is digital.

But, we are still at the beginning of a very exiting journey and we see now even much more potential than at the time we started in 2013 on this Process Mining happy path!

Bosch: Process Mining—A Corporate Consulting Perspective

16

Christian Buhrmann

Abstract

As a large corporation and global player, Bosch has several Process Mining use cases implemented, in different business divisions, across different continents, and in various process types, e.g. P2P, O2C, production, and ticketing. The Bosch Process Mining setup can be characterized as a top-management-driven central approach. Planning and execution is steered by a cross-divisional team consisting of the in-house consultancy, central IT, and divisional coordinators of the participating business units.

Challenge

As a large corporation and global player, Bosch is dealing with a heterogeneous, historically grown organizational structure and hence a large variety of business processes. Following the trend of a stronger focus on digitalization, the process landscape is becoming more digital. This leads to the creation and processing of digital footprints and a massive increase in the available data volume. Traditional and paper-based analysis and optimization tools reach their limits and provide only unsatisfactory insights in the big data context. Therefore, there is a high demand for more powerful and effective analytical tools. Deep process understanding is a key component when improving and optimizing corporate processes. In this context, the handling and analysis of process data is a valuable asset to ensure future competitiveness. Considering this, the use of tools such as Process Mining is a logical consequence. However, there are several constraints when it comes to implementation.

C. Buhrmann (✉)
Bosch AG, Bosch Management Consulting, Stuttgart, Germany
e-mail: Christian.buhrmann@de.bosch.com

© Springer Nature Switzerland AG 2020
L. Reinkemeyer (ed.), *Process Mining in Action*,
https://doi.org/10.1007/978-3-030-40172-6_16

To a certain extent, it is necessary to adapt the respective software to specific characteristics of different business units and process types. Complexity and a high number of various processes require high process expertise to ensure adaptations.

Major implementation objectives can be summarized by the following goals:

- The initial goal is to achieve full process transparency and to unveil the As-Is processes. Transparency enables the reduction of process variants to achieve further standardization. Eliminating obsolete/unnecessary process variants supports process improvements.
- The key goal of an implementation is the removal of overall process inefficiencies and a corresponding reduction in rework.
- Furthermore, Process Mining aims to discover automation potentials.

Manual tools for evaluating process performance based on process performance indicators and the respective data gathering require a high amount of effort, time, and costs. Enormous amounts of data make it difficult for traditional optimization techniques to identify hidden inefficiencies or reveal unknown/unexplored potentials. In consequence, the traditional process modeling techniques reach their limits, as they do not guarantee full transparency. Using additional tools to support or substitute former paper-based methods helps to achieve the goals.

Profound methods must be IT based in order to capture recursions and unnecessary manual effort. By using Process Mining, processes can be visualized and structured based on digital footprints derived by the respective IT source system. Before starting the implementation process and roll-out, a variety of challenges and obstacles may arise. Often coordinative effort and diverse interaction with the respective participants is required. Additionally, change management and potential resistance within established structures needs high attention. At the beginning, setbacks may arise, while introducing Process Mining into corporate structures, as expectations might be high and quick results may be expected.

During the implementation of the first pilots, significant improvements were often revealed. Nevertheless, implementation and expansion of Process Mining may be more complicated and complex than initially expected. This is often related to the need for high process expertise on a detailed granularity level. A key factor is the cooperation and coordination of those involved.

Use Case

The Bosch Process Mining setup is based on a top-management-driven central approach. The planning and execution is the responsibility of a cross-divisional team consisting of the in-house consultancy, the central IT, and divisional coordinators of the participating business unit. The implementation scope includes all processes that contain digital footprints in IT systems. This leads to a diverse relevant process landscape from different divisions. Besides various business units, different process types, such as P2P, O2C, production, and ticketing, are covered. A worldwide roll-out can include multiple countries. In order to manage the complexity, participants/users are trained and the roll-out has to be carried out in an efficient manner. Bosch follows a specific and standardized method, which includes six project phases:

- The first phase is starting with the preparation, which consists of setting up the organizational communication, staffing, and overall project. At the end of the first phase, the operational part starts.
- The second phase includes the scoping. The process scope is defined as well as the specific use case and ideas are generated regarding possible dashboards. Workshops are conducted in order to gain additional information and elaborate on various concepts.
- The third phase is focusing on tool introduction and infrastructural setup. This phase contains the implementation of the software and data by the central IT department. Users are trained according to their assigned roles (e.g., Data Viewer, Data Analyst, and Data Scientist). After testing and revising the developed dashboards, the particular use case goes live.
- The fourth phase consists of first analyses. The As-Is and the To-Be processes are compared and potential root causes for deviations are derived. Lastly, an implementation plan is designed/drafted in order to tackle potentials.
- The fifth phase is focusing on the implementation. Improvement measures, which were identified during the analysis, are implemented.
- The last phase contains further analyses, the full implementation of the identified improvement measures and the documentation and tracking the impact of the improvements. This phase is called the sustainability phase.

Impact

The achievement of full process transparency from a general overview to a detailed perspective on single activities is one of the most important impacts. The visibility of all process variants and the comparison of the As-Is with the To-Be process provides savings potentials and efficiency gains. Generally, a process can be analyzed regarding several process dimensions such as the degree of automation, efficiency, speed, and standardization. Within these process dimensions, various Process Performance Indicators (PPIs) can be evaluated:

- *Automation:* For instance, if the aim is to increase the degree of automation, then a potential PPI can be, e.g., the automation rate or EDI rate.
- *Efficiency:* If the aim is to increase efficiency, the process analysis should focus on quality and waste. Therefore, typical PPIs focus on, e.g., change activities, iteration loops, and blocks.

- *Speed:* If the aim is to increase speed, then an important PPI can be the throughput time.
- *Standardization:* If the aim is to increase standardization, the process analysis should focus on harmonization, accuracy, and reliability. This approach mainly addresses the quantity of process variants.

Briefly, the major impact Process Mining provides is full process transparency and the corresponding potential improvement opportunities. For instance, the existence of certain manual process steps is visual and related root cause are unveiled. Consequently, the impact is represented by the revelation of potentials, the discovery of process inefficiencies, the execution of conformity checks, and process optimization.

Technology

The technology infrastructure regarding Process Mining consists of a frontend and a backend architecture. The PPIs and the process analysis of the frontend dashboards are strongly adapted to the corresponding business unit needs. Generally, data analysts are responsible for creating dashboards. Data viewers can filter and analyze data. This separation is related to the defined user roles distribution of the Process Mining tool.

In the backend, the main target is to achieve a high degree of standardization and reusability between the different use cases in order to keep implementation costs, effort, and time as low as possible.

Lessons Learned

The important requirements for a smooth implementation are the provision of an adequate and available capacity regarding hardware, personnel, and expertise. Experience shows that the desired roll-out intensity quickly speeds up when improvement potentials are discovered and acknowledged in the respective areas. Accordingly, the scaling capacities should be considered and prepared in parallel. The first use case should be in an area with clear existing process pain points and key resources with process knowledge available. This assures measurable improvements with value add for end users.

With additional trainings and workshops, the knowledge in the roll-out area is deepened in order to analyze more complex use cases. In addition, a continuous sharing of best practices, lessons learned, and the transfer of know-how is a key driver to achieving a successful roll-out. A recommendation is to set up a global regular exchange session at least once a month. Moreover, clear allocation of responsibilities is necessary to ensure a fast and efficient project implementation.

   The establishment of application ownership at the in-house consulting department in combination with the top-management-driven approach secures sufficient organizational focus for the worldwide implementation.

Outlook

For the future of Process Mining, different developments are expectable. The combination of Process Mining and RPA seems to be a beneficial and logic approach to tackle further inefficiencies and unveil further potential savings. Discovering inefficiencies and evaluating them in respect to potentials for automation is essential, as the automation of inefficient processes is not purposeful. The expectation is that the number of users within the company continuously increase, so that Process Mining can be used on a daily basis for continuous process improvement.

   With regard to in-house consultancy, Process Mining will be a valued tool within the consulting toolbox. Process Mining can help to increase accuracy and speed of necessary process analyses.

   Overall, the relevance of Process Mining will further increase as entry barriers are continuously lowered and integration of the Process Mining tool into existing IT landscapes is continuously increasing.

Schukat: Process Mining Enables Schukat Electronic to Reinvent Itself

17

Georg Schukat

Abstract

As a medium-sized distributor for electronic components, Schukat electronic must leverage technological innovation to stay competitive. Operational efficiency is crucial to stay competitive and secure a position as distributor in the value chain. For a better understanding of actual processes, Process Mining was deployed to gain transparency regarding actual order processing. Process Mining providing unprecedented transparency and insights, and Schukat is now amidst a data-driven, continuously changing process which has impact on the whole organization.

Challenge

As a supplier of mostly commodity electronic components to manufacturers of electronic devices, we are under a permanent challenge to offer at highly competitive prices and maintain excellent customer service. If the services, which we provide for the products we sell, add extra value to our customers, the buyers' total cost of ownership can be lowered. This may turn into a competitive benefit compared to buying the same product from a competitor, even if the competitor offers the product cheaper but provides a worse service. On the other hand, our supply chain must be evaluated similarly. We intend not to buy from the cheapest sources but from the "best" provider. Best means in this case competitive prices as well as reliable services, which enable us to deliver our services at the level we promise.

Besides improving the foundation of our business model which is "Buy-Hold-Sell," we must, we want, and we will introduce new or extended business models

G. Schukat (✉)
Schukat Electronic, Monheim am Rhein, Germany
e-mail: Georg.schukat@schukat.com

© Springer Nature Switzerland AG 2020
L. Reinkemeyer (ed.), *Process Mining in Action*,
https://doi.org/10.1007/978-3-030-40172-6_17

based on new services, e.g., supporting the Build to Order (BTO) production approach of our suppliers. And there are many more options to consider.

As an initial challenge, we had to spend significant effort to gain insights into our order and delivery handling processes by extending our data warehouse. Both extracting and transforming the relevant data as well as developing informative dashboards were initially difficult. And even then we did not succeed at a sufficient level to understand how we performed and which critical bottlenecks or false setups were forcing us to use extra resources to stay competitive. As a result, we started to search for alternative options rather than continuing with standard business intelligence methods. During this search we learned that Process Mining and analytics would be the right tool set to reinvent ourselves.

Use Case

We started with several Process Mining Proofs of Concept (PoCs) and chose to implement those which were ranked highest by expected value. Based on this experience, we can look back at a valuable and profound Process Mining knowledge in the context of process management.

In the following, we present three selected use cases which have yielded the highest impact. They represent different approaches to process management best practices.

- Bottom-Up—Analyze first
- Top-Down—Design first
- Insight to action—Trigger live action

Bottom-Up—Analyze First

In this use case, we wanted to know how we were performing in the order and warehouse picking process. Our same day shipment rate was below target and orders which we received in the afternoon were too often only ready for shipment on the next day. Furthermore, an internal survey revealed that the staff had to do too many routine tasks manually, as the automation rate was too low.

To understand what tasks have been performed, how they were conducted and in which time sequence they happened for an implemented e2e process, insights are the crucial foundation for change. In our approach we established the following standard methodology:

- Extract and collect relevant event data from all involved systems and transfer this data into the Process Mining system.
- Develop Process Mining Analytics which address the defined issues
- Identify the major causes for critical issues.
- Redesign the existing process by adapting the Business Process (BP) model of the existing process remove these issues.

- Implement the new process modifications.
- Loop.

As an example, we deployed the approach for our order to cash process, which allowed us to find among others the following important causes: by offering free text entry fields in our web order forms to our customers we had to manually adapt the entered orders when they had been transferred into our ERP system, resulting in too many cases with high additional effort. Credit limit and other checks on customer orders were too restrictive, which led to significant numbers of orders held up with a delivery block. New insights showed that only less than 20% of all web orders could be handled automatically. And the racks filling degree in our warehouse had increased year by year, in parallel to sales and revenue growth, causing an unexpectedly high effort within the picking process.

To improve our process efficiency, we extended the web order form for most of the typical customer requests, which had been previously entered into the free text fields. This was possible with specific fields in the web order form which can be transferred directly without manual intervention. In addition, we eased and differentiated order review directives so that critical orders are still temporarily blocked to be checked and confirmed by sales staff. Most other orders can be processed without block and thus without additional effort.

The extra efforts resulting from confined space, as a consequence of the limited warehouse capacity, could only be resolved by expanding the capacity, which we did by building a new warehouse.

Top-Down—Design First

For implementing new business processes top down, we have defined the following new methodology, which has been implemented companywide:

- Transform the business goals into process objectives
- Design chains of processes with BP models which cover all aspects of the new service or business model.
- Implement the new processes in hardware, software, and handling instructions according to the BP Models
- Execute
- Analyze the new process chains with Process Mining tools
- Derive change requirements from process analytics discoveries
- Collect change requirements from all stakeholders
- Loop

We applied this methodology to architect the logistics layout and configuration for our new automated warehouse. The guideline for blueprinting contained four major goals: (1) enough capacity for a business volume which had been projected for

2030, (2) high automation rate, (3) sufficient spare space for new business models and further growth, and (4) high flexibility for agile changes.

The business process blueprint was made available as reference for all implementation partners, hardware, machine control, and software. The partners translated the blueprint into their kind of plans and implemented accordingly. Our intention was to have all relevant data for detailed Process Mining available with the new build automated warehouse from day one after go live. Due to data access, transfer, and conversion problems we had a holdup. Finally, for the technical installation, delivery data from the ERP and more detailed delivery data from the Warehouse Management (WM) systems, together with commissioning log and delivery processing data from the machine controllers, is collected and consolidated. All relevant events and their timestamps are extracted—bracketed by delivery line number as case-key—and transferred into the process analysis and mining system.

As a next step, the insights gained with Process Mining for the new warehouse are used to optimize and fine-tune its configuration and subsequently the operational processes. The delivery process manager evaluates and ranks further enhancements submitted by either the involved stakeholders or him. An optimized process layout is designed in BPM. The new layout is implemented according to the new or modified model.

Loop

Generally speaking, these best practices implement a processes and process chains factory.

Trigger Live Action

- Define sets of rules for event traces and process step times which describe a setting which should be avoided.
- Set up a detection method within the Process Mining system which will trigger a signal to users or operational systems
- Act on the signal to solve the issue

It is common knowledge that best designed processes and automation are never perfect. Even with high efforts to make it nearly perfect, we will still have issues and failures in the supply chains, inbound, internal, and outbound. Not often, but neither neglectable. We have to cope with some cases where early intervention is necessary to avoid failed or delayed deliveries or other kinds of services worse than what the customer expects.

The typical failure symptoms can be identified and described. Based on that knowledge we have implemented several action triggering settings. For example, if special kinds of orders or deliveries get stuck, a signal is sent by the action engine to those users who normally take care of these cases. Interventions to solve the issue causing the holdup are initiated. Delivery processing continues as early as possible. A similar setup is implemented for first signs of supplier's delivery failures.

Impact

The reinvention of our operational processes allowed us to achieve significant impact: The rate of orders which can be directly delivered without manual approval increased from below 20% to more than 60%. And the positive trend continues, as we continue improving.

Delivery processing times have been more than halved due to well-designed processes supported by a high automation rate in the new build warehouse. As a major impact, we are now able to ship all standard orders at the same day or exactly at the time requested by our customers.

The rate of special-order delayed shipments caused by nonnormal process steps went down to nearly nil.

Technology

The vast majority of our business process core data can be extracted from standard database tables stored in an SAP R3 ECC system, running on SAP HANA. Further sources of process data are our SAP EWM (Enterprise Warehouse Management) system and the KNAPP KiSoft machine control system for our automated warehouse. Together with some more process event data from other subsystems all this data is transferred, with a consolidation layer in between, to our central Process Mining system.

Live analytics is done by the Action Engine (AE) running in an Intelligent Business Cloud. The AE is connected by encrypted web services to our Process Mining System on premise. Triggers for actions, e.g., workflows, are similarly sent from the AE to our on-premise transactional systems. Schukat electronic is a co-innovation partner for the AE.

Lessons Learned

In retrospect, we should have focused earlier on a well-defined and high efficient change management setup. If you are not prepared for a high rate continuous change, you will not gain much from many insights of mining processes. Possible improvements will stagnate.

With respect to the "order to cash" or even the lead to offer to order to cash process steps is looking only at a subset of all aspects of the full customer journey. We recognized that we are missing major interactions of our customers with our online and personal customer services before order and after payment. Activities like web interaction, inquiries, extra offers, returns, warranty claims, etc., are at this stage not contained in our implementation of the order handling Process Mining. In this respect we are still partly blind and may miss important aspects of the full e2e processes. We cannot be sure to always come to the right conclusions and so might choose inapplicable changes.

As a basis to analyze a flow of processes and connect it to an e2e process, a case key common to all processes and subprocesses within the chain is required. With every extension to widen the scope the process chain gets longer and it becomes more difficult to find a case key covering all different activities and variants, e.g., looking at the process from customer offer to order to cash. As a case key, we have defined invoice number and line number. A union of all events backward and forward connected to that invoice allows us to include the invoice itself, the delivery note, the delivery, the order confirmation, the order, and, in case the order is based on an offer, this as well together with the payment into the event table. While this insight is almost complete, we miss all the offers which were not converted to an order. In consequence, this data model does not allow us to analyze how and where differences exist between successful and failed offers. Similar deficits in this data model exist for cancelled order lines.

In general we will try to extend every data model to an as-broad-as-possible e2e scope, with all kinds of detailed events, and in depth.

On the other hand, if your data model contains too many details, too many events, too many subprocesses, you might come to a situation where you will not see the wood for the trees. We still do not know how to fully measure, analyze, and understand the impact of repeated subprocesses within the O2C process, e.g., several partial and split deliveries for one order line caused by temporary low stock.

Process Mining, Process Design, Process Management, and the Action Engine have become our toolset to reinvent ourselves. By combining these tools we are now able to adapt and change our processes and services agile, fast and keep the risk of losing customers due to initial bad experience low. The buildup of the new automated warehouse which is the successor to a computerized, but still run by only manual handling processes, warehouse. We reinvented the way we work e2e.

Architecting the process implementation and change toolset is reinventing ourselves at a higher level, we can now design and develop our data-driven approach.

Outlook

For future improvements, it might be interesting to enrich Process Mining data with customer experience data collected by customer feedback tools. An overarching customer journey view would ideally include all activities resulting from customer touchpoints. Such a future framework would enable us to include the most important external evaluation level into our process analysis results.

The principle to mine a newly designed process or service immediately after going into production can be carried further. The mining should be well engineered and directly connected to action triggers for critical settings. Thus it could become a good foundation for agile go lives and agile process optimization. Consequently one would be at a substantially lower risk caused by unforeseen critical issues within the new process or service. An early agile go live with minimal viable processes or services can be ventured with relatively sufficient safety. The Action Engine becomes a safety net for new implementations and major changes. Furthermore early Process Mining provides a good basis to avoid unnecessary inefficiencies right from the beginning and so can accelerate further advancement.

Many event logs which are the base data for Process Mining only contain event finished timestamps. With more detailed event data like event starting timestamps, queue waiting times, used resources, and added features and functions in the Process Mining system, we might be able to build a reliable process cost calculation. Dicing and slicing process costs, e.g., for O2C costs compared with the marginal returns for the same grouping, could help to identify those materials, material groups, customers, and customer groups, etc., where we earn most or least.

This kind of analysis can be carried further. Many organizations are nowadays orientated toward minimizing their negative impact on nature and the environment by reducing their energy and resource consumptions. A Process Mining system able to do process cost calculations could do energy and resource calculations as well, either based on direct measurements by IoT devices or on external data. Thus it will enable us to get a differentiated and detailed view on the amounts of consumption for every process step. When having very broad process models, far-reaching in- and outbound in the supply chain, we can attribute and balance consumptions. These insights will be an optimal, facts-based foundation for actions to reduce them.

We expect that in the near future Process Mining might become the most important analytics tool for day-to-day business analysis. Standard BI tools will become less important for managing companies.

To maximize the positive effects of Process Mining, major process design and process management restructuring within the organization is needed. It is recommended that the organization set up a change management system. Changes within e2e processes often generate impact across department boundaries. In addition to a change management system, the role of a process owner with its responsibilities, its decision rights, and its position within the organization's hierarchy must be properly defined.In consequence, a management position of a Chief Process Officer[1] could be established. In the past such a role definition was more focused on process design and management to achieve the goal of process excellence. Now with modern Process Mining tools it might become a lot more powerful role with a much higher impact within the organization.

---

[1]Example: https://www.processexcellencenetwork.com/business-transformation/interviews/time-for-high-value-process-excellence-interview-w

Links

Schukat   Imagefilm: https://www.youtube.com/watch?v=v8R01AVZrj8&
    list=PL9UMfh3sw0BTm8ueMSmvciKY6pvdISxjK
Schukat Insights "Transparenz": https://www.youtube.com/watch?v=Gyt3iUCtyrs

Siemens Healthineers: Process Mining as an Innovation Driver in Product Management

18

Jutta Reindler

Abstract

Process Mining applications were adopted in the already productive business intelligence platform to support the constantly developing Computed Tomography (CT) product and service portfolio. The approach was driven by innovation management, recognizing the unique opportunity to optimize the CT product design and software workflow based on the real interaction between human and machine. For typical questions such as "how performant and user-friendly are the CT devices in clinical routine?", "are programmed/predefined workflows accepted?" or "does the new innovative tablet control improve patient workflow?" Process Mining provides unprecedented transparency and thus the basis for strategic improvement.

Challenge

Anyone who has ever been examined with the help of a larger medical device knows them: the static computer workstations (consoles) with which they are operated on from a separate room at a distance.

Siemens Healthineers has developed a flexible and modern alternative solution for these static workstations: a mobile tablet control system. Currently, both technologies are in use—and here was the challenge: an aggregated view on the relevant workflow steps for a basic CT examination, including the classical workstation activities as well as the activities from the new mobile components like tablet or new features like a camera.

The operating software for a basic examination from patient registration to image acquisition consists of two million lines of code and gives an idea of the complexity

J. Reindler (✉)
Siemens Healthineers, Erlangen, Germany
e-mail: Jutta.reindler@siemens-healthineers.com

© Springer Nature Switzerland AG 2020
L. Reinkemeyer (ed.), *Process Mining in Action*,
https://doi.org/10.1007/978-3-030-40172-6_18

143

of the entire workflow programming. The system event logs are generated in unstructured format, which is a typical problem for IoT data. The system events for the new CT system platform are specified and created by different, globally distributed feature teams and components. For example, the new mobile software was designed and programmed by a different feature team than the traditional console software event, though the event triggers the same action. Besides creating a short-term transparency on the workflow execution, the overarching goal was to standardize event logging based on the learnings of Process Mining.

In addition to these technical challenges, there were structural and intercultural hurdles. For example, the development authority (R&D) for the newest CT platform resides in China, while the responsibility for individual components also resides in India. All questions related to structure, consistency, and completeness of the data had to be aligned between these colleagues. These challenges are reflected in the app intro page shown in Fig. 18.1, with tables presenting the number of incomplete examination and the age of the data on activity type level (due to the complex data chain).

In addition to this fundamental goal of providing a valid and robust database of IoT data to enable plausible and reliable analysis of various CT devices, it was intended to discover real user behavior and variants in the application processes and provide answers to the following questions:

- How do our customers actually work with our CT devices?
- In which parts of which workflows do errors occur?
- Which workflow variants make up the customer's clinical routine?

Fig. 18.1 App intro dashboard for CT usage data

Use Case

In cooperation with the consultants of the Process Mining solution provider, the pilot project first worked out whether and how Process Mining can be applied to the CT scanner data and how the existing data with all the special features mentioned above (unstructured, nonstandardized, etc.) must be prepared and standardized. A major challenge here, again complicated by the international distributed development teams, was the correct definition of the elementary Process Mining elements as a *unique identifier* for a process or subprocess. For example, the project team found that the term *key* and its requirements could be interpreted quite differently.

The basic analysis included 24 different activities shown in Fig. 18.2, from patient registration to reconstruction of acquired CT images with the tablet and camera usage events as separate activities.

In order to create a better internal acceptance of the new analysis apps, the Process Mining applications were rebranded to "workflow mining" at the suggestion of a

Fig. 18.2 Activities of the basic analysis

2D Camera

Bolus aorta ROI

Button

Button_Confirm

Button_Finalize

Button_GO

Button_Next

Button_Ok

Cardiac heart cycle

Confirm planning

Confirm position

End

Examination end

Examination start

Fast planning

Parameter changed

Protocol load

Recon application

Registration

Scan

Start

System message

Topo

Chinese data scientist. The aim was to get rid of the wrong association to the main Process Mining use cases in regards to business processes.

The dashboard in Fig. 18.3 provides an overview of the specific system setup, the different process variants, number of exams per day, examination cycle time, and organ-specific spread that shows the clinical use of the system.

Minor workarounds at the database level and creative solutions, e.g., for defining and extracting meaningful time stamps and logging structures, were implemented quickly. Fortunately, thanks to the associative technology of the BI-based Process Mining solution and the close collaboration with the Process Mining Consulting team, data quality issues were quickly resolved. To reduce, e.g., the extraordinarily high number of process variants, all variants happening only once or with incomplete basic workflow (e.g., no scan) were excluded from the visualization.

From raw to ready: The question "can we apply Process Mining to our scanner event logs?" was answered with a "Yes." In a very short time, user-friendly Process Mining apps were created, rolled out, and independently extended by Siemens Healthineers.

Today, the most common workflow variants can be immediately filtered to visualize the customer's clinical routine by using Process Mining, as shown in Fig. 18.4.

Inspired by the type of visualization, software architects asked how many different protocols make up 80% of the daily workload of a system. With the seamless BI integration, a pareto analysis was the solution preferred, as shown in Fig. 18.5. The idea was to load the most used protocols in memory.

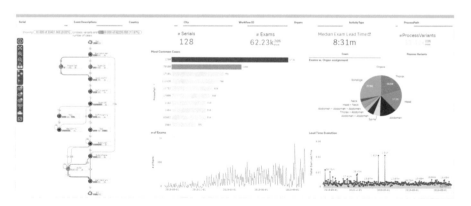

Fig. 18.3 BI-based Process Mining dashboard with process analyzer (left) and process performance indicators (right)

Fig. 18.4 Detailed view of the most common cases

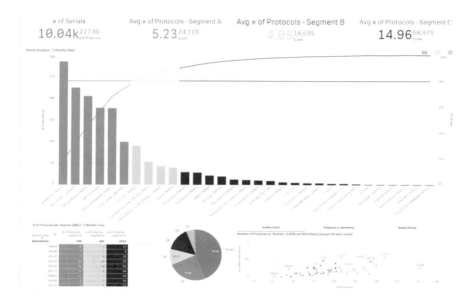

Fig. 18.5 BI-based Process Mining dashboard: Pareto analysis

Impact

As one CT collaboration manager in the United States exclaimed: "Stopwatches and cameras GOODBYE!" With Process Mining, collaborative customer insights can be more easily gained by just clicking on the relevant apps and seeing 100% of the Process Variants (Fig. 18.6).

First, Process Mining supports internal efforts to standardize the CT workflow based on real customer usage scenarios. Process variants could be uncovered and thus the basis for standardization measures could be created. In CT's experienced-

Fig. 18.6 Detailed view on
process performance indicator
"Process Variants"

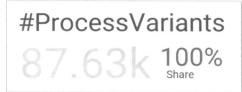

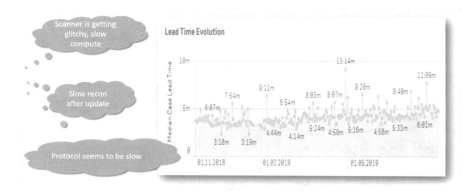

Fig. 18.7 Detailed view on process performance indicator "Lead Time Evolution"

based testing labs, the data supports real usage scenarios, not predefined static test
cases. Even activity outliers, like customers leaving patient tabs open and
complaining about respective system timeouts, can be analyzed and quantified in
respect to relevance.

.Lead time analyses—especially in combination with service ticket information—
show a clear performance profile for each system (Fig. 18.7). Performance
requirements are customer perception critical when it comes to the right setup of
computer hardware.

It is also easy to see whether a customer is becoming faster in interacting with the
system, i.e., in the more complex needle-guided CT interventional scenarios, where
the doctor has to use footswitch and "joysticks" to perform the exam. The lead time
evolution is shown in Fig. 18.8.

Another interesting aspect is the comparison of automated vs. manually adjusted
workflows (Fig. 18.9). With a sufficiently large number of examinations, it could be
proven that the throughput time is significantly reduced by letting the systems set the
parameters.

Siemens Healthineers can generate a competitive advantage for itself by
implementing fast learning cycles to improve the operating software not only by
interviewing customers and getting feedback from the regional sales and service

Lead Time Evolution

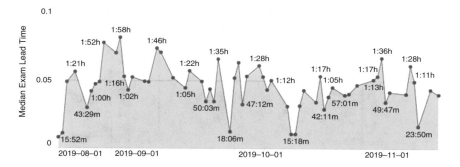

Fig. 18.8   Detailed view: a system with a learning curve when performing interventional exams

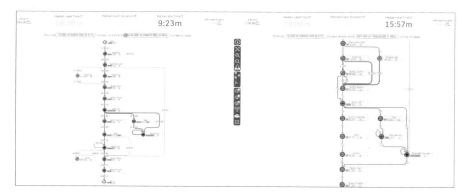

Fig. 18.9   Process Mining comparison dashboard: comparison of cycle time for automated exam workflow (left) vs. manual adjusted parameters (right)

specialist, but this can be backed up with this additional data and insights on a broad scale.

Process Mining insights can be linked to already existing data classes, countries comparisons as shown in Fig. 18.10, large accounts, or dedicated customer segments with specific needs. The knowledge gained then flows directly into an improved product design.

At Siemens Healthineers, the Process Mining technology is a valuable investment with a strategic vision. Jutta Reindler, the driving force and the innovative spirit behind this strategic deployment scenario, notes: "When will it ever happen that the use of software can generate a real competitive advantage? Process Mining analytically supports us in identifying issues and customer behavior faster and accelerating our learning cycle. That's great."

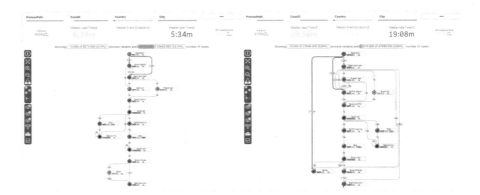

Fig. 18.10 Process Mining comparison dashboard: average exam duration for high-performance country China (left) vs. diversified usage scenarios in the United States (right)

Process Mining Enhanced by New Types of Visualizations

The focus was and is not primarily on analyzing classic key and/or process performance indicators, but on being able to design better products and CT systems in the long term through data transparency: on the basis of real customer and user experience.

Process Mining charts have thus been enhanced by several new visualizations.

Exam Throughput per Hour

A colleague asked an exciting question about the patient throughput per hour. Thanks to the symbiosis of Process Mining and associative business intelligence, a corresponding analysis with an appealing and easily interpretable heatmap was created without great effort. This could also be used to answer further questions arising from this, e.g., whether there are accumulations on certain days or at certain hours (Mondays or mornings, etc.) (Fig. 18.11).

Trending of Examination Times

To better understand the customer's perception when using the system, the examination times can be not only visualized in a line chart but also trended. The trending indicator (in our case the linear regression) can show as a result, e.g., a learning curve in case the examination time is reduced over a long period of time, a stable performance or an increase in examination time in case the customer has problems (e.g., with the hardware).

Figure 18.12 shows the development of examination times for a specific site with a linear regression line on top.

Fig. 18.11 Detailed view of heat map "Patient Throughput"

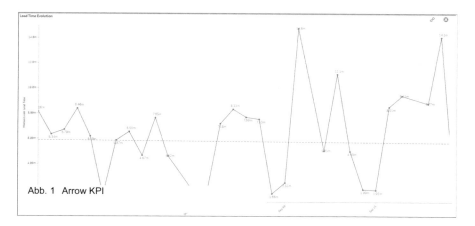

Abb. 1 Arrow KPI

Fig. 18.12 Detailed view: including linear regression in exam lead time evolution

The regression functionalities are an integrated part of the BI software solution and can be included in the visualization or as a KPI by itself. Represented by a red/green/grey arrow, the KPI can be used to sort fast and thus quickly identify the respective problem sites, as shown in Fig. 18.13.

Visualization of Individual System Settings

In addition to the standard Process Mining visualizations, attractive word cloud visualizations or Sankey diagrams are also used to provide the user with different views of the system-specific configurations in a visually appealing form as in Fig. 18.14.

Serial ID 🔍	Trend		Avg
75438	▼		‖‖‖‖‖ (37%)
75936	▼		‖ (15%)
105077	▼		‖ (9%)
75946	▲		‖‖‖‖‖‖‖‖‖‖‖‖‖ (105%)
75947	▲		‖‖‖‖‖‖‖‖‖‖‖ (93%)
122012	▲		‖‖‖‖‖‖‖‖‖‖‖ (90%)
122002	▲		‖‖‖‖‖‖‖‖‖‖ (89%)
122028	▲		‖‖‖‖‖‖‖‖‖ (84%)
75967	▲		‖‖‖‖‖‖‖‖ (69%)
105103	▲		‖‖‖‖‖‖ (48%)

Fig. 18.13 Detailed view: trending performance for multiple systems

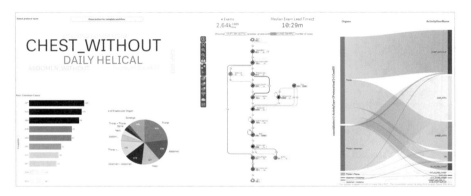

Fig. 18.14 BI meets Process Mining: Word cloud and Sankey diagrams

Technology

In the following, the already mentioned technical challenges with regards to data preparation will be examined in more detail and the BI-based Process Mining technology will be presented.

Event Log and Data Architecture

The retrieval of the CT Workflow Event Logs used for Process Mining was not trivial, as the databases are distributed worldwide. For example, the data from the Shanghai databases had to be copied to a local server, because the response time was too short to be included directly in the event log creation process. Intermediate copying to a faster local server takes place on a daily basis. The CT log data was extended by relevant context information and supplemented by SAP data, so that a meaningful connection of different source systems was realized. Problems such as a missing connection between the user interaction event log and the actual case ID could be creatively solved and valuable information could be reconstructed.

The data architecture and its quality play a key role for later data and process analysis based on Process Mining algorithms. It is therefore not enough to find the logic for event log generation solely in the SQL database and transfer it one-to-one to the Process Mining Tool—even if Siemens Healthineers wanted to be able to perform data preparation, event log generation, and Process Mining in one tool. Once the event log has been created, however, there are actually no more system breaks. The Process Mining algorithms transform the event log in the same tool in which the data modeling for the data analysis and the data analysis itself take place.

Process Mining App

When the CT Workflow Mining app was first created, requirements changed daily, almost every hour—whenever the think tank held its meetings. Thanks to the flexible BI basis, the changing requirements could always be implemented in the shortest possible time. The BI platform allows you to link the event log with other data sources or to reuse other BI applications. This enabled rapid app development, combined analyses, and the flexible addition of new key figures. Special requirements could also be met without any problems, since both the backend code and the mining algorithms are readable and editable. One modification, e.g., was the improvement of event sorting through defined sort keys.

Hosting

Although the Process Mining solution used is available in the cloud, Siemens Healthineers—as many German companies still like to do—prefers to host the software on premise, as this does not make any difference to users. The powerful in-memory engine enables—equally measures for both deployment variants—powerful filtering and recalculation of charts. Thanks to the in-memory engine, several employees can analyze their extensive CT workflow data in parallel in the same browser-based Process Mining app. The browser-based approach gives every licensed employee access to the published version of a CT workflow mining app—worldwide.

Self-Service Process Mining

The BI-based Process Mining solution enables Siemens Healthineers Self-Service Process Mining. What does this mean? Every analyst can create new worksheets, new diagrams, and new formulas via the frontend. Special functions in the frontend offer international teams collaboration opportunities: insights can be shared or new analyses can be published for colleagues. This is one reason for the creative development that the app has undergone during the first year. Siemens Healthineers has independently developed the template app that was originally made available and had created heat maps, word clouds, and Sankey diagrams to answer use-case-related questions. The original system-type-specific applications were branched to focus on specific usage behaviors, like high throughput customer or interventional sites.

Lessons Learned

Overall, the use of Process Mining technology at Siemens Healthineers was and is a great success. More and more analytical purposes are being recognized and used. In the pilot project, various challenges could be mastered—this was not possible within previous data mining projects.

The key to the success of the Process Mining project was the clear focus on the core CT workflow that had been set for the pilot. The clear focus helped Siemens Healthineers to first get a feel for the software itself as well as for the correct data acquisition, provision, and quality needed. Not all activities need to be included in an application, it very much depends on the use case. Different variants in activities can be linked in the data model as attributes, not separate activity types.

The Process Mining solution was quickly accepted by the users as it was seamlessly integrated into the analytics platform. Activity tables are just an add-on to the current customer, service, quality, and IoT data. This reduces data preparation efforts as well as an easy administration of the dashboards. The flexible and intuitive way to associatively analyze data in well-structured analysis cockpits contributes to the growing acceptance or urge of colleagues to map even more usage scenarios.

Jutta Reindler states, "Process Mining is not something you immediately establish for your IoT data. The use of BI-based Process Mining is a new step towards visibility, which I never envisioned before. We've done typical data exploration with tables, line and pie charts for quite some time."

The biggest hurdles and most time-consuming milestones, some of which have already been mentioned before, are outlined in the following tables.

Challenge of Gathering Data

Challenge	Description and solution
Definition of a meaningful Process Mining key	A different or unclear understanding of the term "key" in the context of Process Mining led to a project team from China initially supplying other/false identifiers. After clarification of the exact terminology/function/requirement for "uniqueness," this hurdle could be overcome.
Events without identifier	There were activity types in the data that did not have an identifier. This was solved by a workaround at the database level to add the workflow key based on timestamps.
No unique keys	The problem of nonunique keys was solved by a workaround using additional timestamps/logging structures, i.e., in principle by creatively completing data. The associative technology allowed quick uncovering of data quality issues.

Structural and Organizational Challenges

Challenge	Description and solution
Heterogeneous, international development teams and unstructured IoT data	Those developers who developed the new mobile tablets did not develop the consoles as well → this means that the events have different names and the consoles have different processes/workflows → the process variants differ → the number of variants increases. Additional parameters were and are used to work on an aggregated view of the process data of the console and tablet workflows that differ greatly (from a data point of view).
License procurement on the basis of central IT	For the BI innovation team, permanent funding for license procurement was a major hurdle or a slowing milestone from a time perspective. The breakthrough was the impressive prototype that helped to convince the BI platform team to establish this extension on a greater company level.
Training sessions	Another time-consuming component of the project was the subject of "training courses." Project success and employee feedback, however, spoke in favor of this time investment.

Outlook

Following the successful pilot project, an international roll-out was initiated, so that the new opportunity for process analysis provides valuable insights for all employees with corresponding needs. But there were also ideas for new areas of application during the pilot. These came both from the project team itself and from outside.

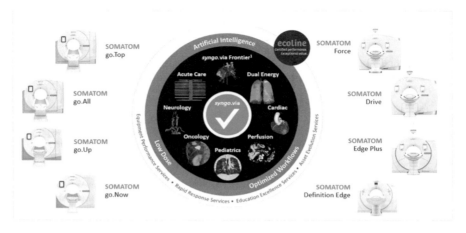

Fig. 18.15 SHS Computed Tomography portfolio

Voices were quickly raised that Process Mining could also be used for other purposes like the tracking of R&D backlog or the offer to order process.

Process Mining apps could be part of the Siemens Healthineers remote service support if the data chain supports real-time scenarios.

Following the example of scanner data, Process Mining can of course also be used in the future to optimize other products from the portfolio—also in the area of innovation management. The thoughts of further establishing and using Process Mining in the future are therefore going in various directions.

Links and Further Reading

Siemens Healthineers: https://www.siemens-healthineers.com/

Introducing groundbreaking products and services for decades, Siemens Healthineers is known as the innovation leader. This strategic approach is backed up by dedicated innovation budgets, crowd funding projects and career paths as "key expert." Jutta Reindler is Innovation Manager for "Business Intelligence" of Computed Tomography at Siemens Healthineers. Siemens Healthineers enables healthcare providers worldwide to increase value by empowering them on their journey towards expanding precision medicine, transforming care delivery, improving patient experience, and digitalizing healthcare. A leader in medical technology, Siemens Healthineers is constantly innovating its portfolio of products and services in its core areas of diagnostic and therapeutic imaging and in laboratory diagnostics and molecular medicine. Siemens Healthineers is also actively developing its digital health services and enterprise services.

In fiscal 2018, which ended on September 30, 2018, Siemens Healthineers generated a revenue of €13.4 billion and an adjusted profit of €2.3 billion and has about 50,000 employees worldwide. Computed Tomography (CT) is one important area of the diagnostic imaging division, with a currently active installed base of 27,000 systems around the globe.

With the great objective of constantly developing the Computed Tomography product and service portfolio and, as a result, striving to improve patient experience, the exciting application scenario of Process Mining technology described here points in the same direction.

Bayer: Process Mining Supports Digital Transformation in Internal Audit

19

Arno Boenner

Abstract

Internal Audit at Bayer AG was an early adopter of the innovative Process Mining technology though—at first glance—an audit organization does not seem to be predestined to be the typical point of entry. Two factors favored this step: On the one hand, Bayer AG has been using SAP globally in its core processes for many years; this systemic homogeneity is not to be underestimated for the implementation of a Process Mining application. On the other hand, the majority of the audits performed by Internal Audit at Bayer AG, particularly in the commercial area, are strongly process driven. Although proven and extensive table-based toolboxes were available, it is very difficult to describe and interpret a global e2e process using tabular analyses and to audit it in a risk-oriented manner. Intense search discovered visual Process Mining as the perfect solution. In the start-up hall of the international IT-Fair CEBIT in 2012, the first meeting with an innovative young company took place. On the personal wish not to endure PowerPoint presentations, but only to see live data of real existing systems, it quickly became clear that the product came very close to what we had been looking for since many years. Our use case outlines the challenges of implementing Process Mining and how it drove the digital transformation of Internal Audit at Bayer.

Challenge

After the initial euphoria, it became clear that the introduction of Process Mining was not a success in itself and initially only meant one thing: work. The initial product as of 2012 was obviously very different from today's capabilities. For example, the "Content Store," a kind of digital supermarket where prefabricated analyses can be

A. Boenner (✉)
Bayer AG, Leverkusen, Germany
e-mail: Arno.boenner@bayer.com

© Springer Nature Switzerland AG 2020
L. Reinkemeyer (ed.), *Process Mining in Action*,
https://doi.org/10.1007/978-3-030-40172-6_19

taken from the virtual shelf and built into the customer's process model, didn't exist. While until 2012 data mining and data analysis were mainly table-based analytics formats, the timestamp-driven technology presented us initially with major challenges.

New competencies required While SAP knowledge in combination with Audit Command Language (ACL) knowledge was previously important, it quickly became clear that event-log-based process modeling should become an important competence, though it was not yet represented in our ranks. For the young partner, Bayer's assignment was one of the first encounters with the complexity of globally operating SAP systems. As a redundant approach, over a period of several years a table-based ACL Tool box was used in parallel with Process Mining. Mid 2017 it was 100% certain that Process Mining including the system architecture in which it was embedded had achieved a high degree of "system stability," and that sufficient new competencies had been established, the fallback scenario in the form of a table-based ACL toolbox was dispensed with.

Where to start The cooperation with the Partner would be called "agile" in today's language, since it did not involve the implementation of a proven standard software. The work style can be rather characterized with the attributes explorative, pragmatic, and result oriented. As one of the first questions we asked ourselves how to define a starting point. The aspects of process standardization and data availability in combination with the Internal Audit's own areas of expertise were helpful in answering this question. While the process-related degree of standardization at Bayer in the area of accounting/controlling is very high, this does not apply in the same way to Source-to-Pay (S2P) and Order-to-Cash (O2C). The lower level of standardization in the O2C process is due to the various business and logistics models, which only allow process homogenization on a comparatively low degree. Since the S2P area shows a much higher degree of harmonization and the purchasing competence within Internal Audit is excellent, we ventured into a familiar territory with a new technology that was almost unknown to us. The first prototype was ready after only a few months and was tested in October 2012 during an audit in Hong Kong. Many questions had to be answered: does the data match reality, i.e., the SAP source data? How is the system performance with regard to intercontinental use? How usable and user friendly is the application? Luckily, most of the key questions could be answered satisfactorily and nothing stood in the way of further progress. However, it should also be noted that the unique selling point of Process Mining, the "Process Explorer," was initially only usable to a limited extent during audits—but was an eyecatcher during presentations. A successful pilot phase and the formal agreement of a development cooperation were important milestones for a continuous cooperation for further joint progress.

The choice of system architecture While we had initially concentrated on the operative implementation of process flows and dashboards for test purposes, we had left out the important question regarding the choice of system architecture.

Process Mining is operational on SAP HANA or SQL. In our case, the decision was made by the elimination procedure: since SAP HANA was not yet widespread until about 5 years ago, the establishment of a Microsoft SQL server was the means of choice. Disregarding some minor changes, the setup we chose many years ago turned out to be a sustainable solution till today.

Data integrity challenges   While data integrity initially appeared to be less than spectacular, it temporarily put the entire project in a dangerous position of imbalance. The upload of tables from SAP source systems turned out to be not trivial, but rather extremely difficult. Missing tables and therefore incorrectly calculated data cubes were very common and the partial unreliability at least temporarily undermined the most important asset: Trust in the application provided by the auditors. With a bundle of measures we got the quality problem on the database side slowly under control. However, it was not the only problem, as—in parallel— we were confronted with another challenge.

"Performance-optimized programming" is a keyword that probably sounds in the ears of many Process Mining enthusiasts. The calculation of the data cubes took a disproportionately long time and performance optimized programming was a continuous challenge. Too often the delivery times could only be kept very short and there were unfortunately also some cases with delays in data provision. Our favored approach of presenting e2e process chains posed major challenges for the experts of our Partner, but got gradually resolved.

Use Cases

Due to the complexity of many processes, it may make sense to restrict the process data which shall be analyzed prior to Process Mining. Various dashboard components can be used for this purpose. Dashboard components are representations of ERP data, in this case SAP source data in graphical form. Pie charts, donut charts, column charts, etc., can be used in order to make data intuitively accessible. What is displayed in the form of charts depends strongly on the purpose of the analysis. Figure 19.1 shows four examples of an audit-specific, risk-oriented perspective on

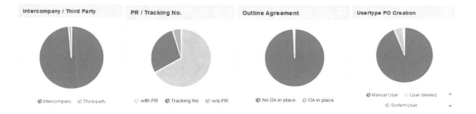

Fig. 19.1 Four Purchase Order dashboard components

SAP data. To enable this type of analysis, OLAP data is stored in a multidimensional database. While a relational database can be thought of as two-dimensional, a multidimensional database considers each data attribute (e.g., Purchase Order (PO) with Purchase w/o Purchase Requisition (PR), PO w/o PR, etc.) as its own dimension. Thus, it is possible to determine an intersection of dimensions from the numerous attributes. The following example from the PO process section shows how an auditor uses the various dashboard components in a risk-oriented way.

Intercompany/Third Party This component specifies whether the transaction is an intercompany transaction within Bayer AG or a transaction with a third party. From a risk perspective, transactions with third parties are classified as riskier and the application allows to drill down deeper into these "third party" transactions.

PR/Tracking No. This component specifies whether the PO is based on a PR or not. The PR ensures that a user has triggered the order.

Outline Agreement This component shows whether a PO refers to an outline agreement or not. An outline agreement is a quantity or value contract created in SAP that contains optimal terms and conditions and is created automatically on the basis of a PR.

Usertype PO Creation In this case we differ between a machine user and a human user. As the human user's activities can be assumed as more error prone, manual transactions, respectively PO creations, are considered riskier than those conducted by machines. Filter and drill downs are provided for the "manual user."

Net Value in LC (Local Currency)/Count per Month The trend in Fig. 19.2 displays two values on a time axis: The number of order transactions (Count) and their value (Net Order Value). The amount of data can be reduced by limiting the time segment.

The above-mentioned dashboard components express Internal Audit's view on SAP data. With risk-orientation and correctness dominating the view. Throughput times, e.g., are not taken into account. The four dashboard components as displayed

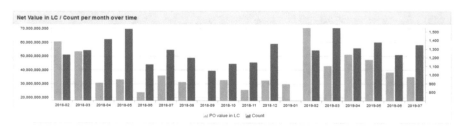

Fig. 19.2 Net value and count per month on a timeline

in Figs. 19.1 and 19.2 represent only a selection and it might make sense to filter on material groups or document types (SAP order types) as well. At the beginning of the analysis, 19,208 POs were available; after risk-oriented application of the various dashboard components, the number of POs was reduced to 251, which corresponds to a reduction of more than 98%. What remains are the POs which

- Represent a transaction with third parties
- Are not related to PR
- Do not refer to SAP Outline Agreements
- Are created manually
- Occurred between February 2018 and July 2019

From an audit perspective, these remaining POs bear an increased risk. With the help of the Process Explorer, the process is examined with regard to deviations (Visual Process Mining, see Fig. 19.3). After the activity "PO Item Creation," two deviations are noticeable: "PO Item Price Decrease" and "PO Item Amount Decrease." To understand the causes of these process steps, the transactional data of the 16 quantity changes and the two price changes must be downloaded to perform root cause analysis using Excel. Process Mining is thus supplemented by data analytics.

These three analysis steps [(1) data reduction through risk-oriented application of dashboard components, (2) visual Process Mining, and (3) tabular analytics] should

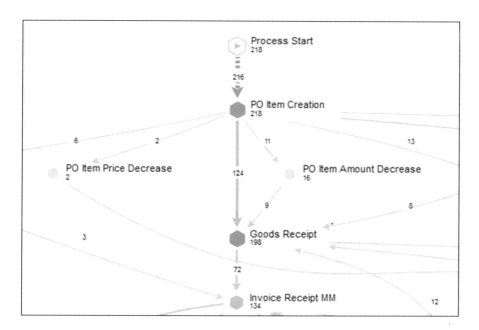

Fig. 19.3 Visual Process Mining for identification of process deviations

take place weeks prior to the actual audit in the country—also known as the fieldwork phase. The auditor is thus able to enter the onsite audit in a very focused manner and, if necessary, identify contact persons in advance or request further detailed documents. By using Process Mining already in the preparation phase of an audit the added value of the audit can possibly be significantly increased.

Impact

With the introduction of Process Mining, a digital transformation from data analytics and data mining to Process Mining was achieved. Mining, understood here as the search for the unknown, was enormously strengthened by the inclusion of event logs. The visual representation of the process enables the detection of process anomalies and inefficiencies. Besides providing transparency, Process Mining also supports root cause analysis. This is exactly where it becomes clear that Process Mining does not supplant data analytics, since tabular analyses can now be used to search for the root causes. A good example is the so-called Touched POs, where Process Mining (Process Explorer) was used to identify numerous orders that were created automatically, but processed manually thereafter. Only the subsequent tabular analysis made it possible to determine the root causes, namely, subsequent price and quantity changes, as well as incorrect master data. It was also possible to identify patterns, respectively approval procedures, that do not comply with Bayer regulations. Many process weaknesses could be identified through Process Mining; the weaknesses could be eliminated through numerous corresponding recommendations. Thus, the internal control system could be sustainably strengthened. In addition, many findings that were not risk related, but had, e.g., a strong focus on efficiency, could be provided for further follow-up to the auditee as well.

Technology

The simplified system architecture is shown in Fig. 19.4. The application used by Internal Audit is currently a local on-site installation that is not hosted in a cloud. In general, a distinction can be made between three levels.

Source Systems These are the SAP source systems that host the data of the approximately 250 legal entities of Bayer AG. The dab:Exporter is installed between the source systems and the database layer, updating movement data such as goods receipts, goods issues, POs, material bookings, etc., once per day. The dab:Exporter is used to update the business data of all entities onto the Microsoft SQL Server. The dab:Exporter has read-only access to the SAP source systems. For reasons of simplification, non-SAP data sources are not considered.

Database Layer This is a central Microsoft SQL server that stores the uploaded data in SQL format.

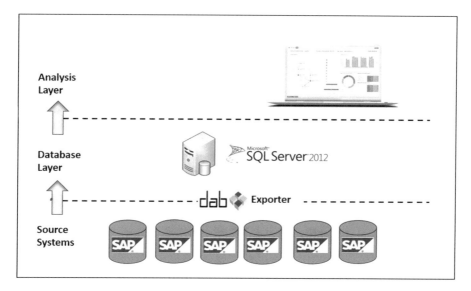

Fig. 19.4 System landscape—simplified Scheme

Analysis Layer   The Analysis Layer is represented by the Process Mining frontend, which is used by the end user/auditor to view the data. Access is possible worldwide via intranet. Authorization management ensures that only persons involved in the audit have access to the company's data for a predefined and limited period of time.

The validation of the data takes place on all three levels and ensures that the data on the three levels are synchronized. Especially when uploading data from the source system layer to the database layer, an RPA application is conceivable as a validation measure to ensure the integrity of the data and will be discussed further below.

Lessons Learned

As lessons learned, the topics of training, coverage, outsourcing, Process Explorer, HR strategy, data privacy, and release changes are explored to share some experiences.

Training   The implementation of Process Mining can be seen as an enormous technical advance. It should be emphasized that the application comes very close to the habits and expectations of IT natives, while the intuitive usability is also to be emphasized. Nevertheless caution is advised: In sober terms, Process Mining is nothing more than a special way of displaying SAP As-Is data, while using Process Mining without SAP process and SAP table knowledge is rather meaningless. To remediate this challenge, the Audit Intelligence team has created a 1-week training course that teaches basics in SAP data and processes and analysis skills, taking place in the form of classroom training at Internal Audit branches worldwide. In order to bring the training as close as possible to the real audit process, exercises in the form of real cases are included, based on real historical data. The training system is available 24/7, 365 days a year.

Coverage Although an activity that carries a timestamp can be formally presented as a process with the help of Process Mining, some areas such as production or HR are largely left out. The reasons are manifold, ranging from poor data quality in the SAP source systems, high sensitivity, or the fact that processes are not mapped in SAP systems and the required competence cannot be maintained within Internal Audit.

Outsourcing Since we have achieved a high level of stability during the last 3 years, we have been able to identify many repetitive activities that we have consequently outsourced to a Bayer Shared Service Center. The tasks assigned to the Shared Service Center are executed in the background and include cutting data cubes and validating the data. This setup allows the Audit Intelligence team to further focus on value adding activities.

Process Explorer As already indicated above, the Process Explorer—although the most prominent product feature to date—is of comparatively little use. As already mentioned, Process Mining can be understood as an As-Is representation of SAP data. From an audit point of view, process deviations are of particular interest. In order to identify these, a normative dimension in the form of a To-Be scenario must be integrated into the process representation in order to determine the delta, i.e., procedural deviations. At present, this type of analysis is based on the knowledge and experience of members of the Audit Intelligence team and auditors, who reflect their experience on the basis of the current representation. Objectification and automation of this analysis step would be very welcome.

HR Strategy A further important conclusion or recommendation is the HR strategy. Within the global Internal Audit organization, a central Audit Intelligence Team has been established at the HQ in Leverkusen, which takes care of the data supply for all audits worldwide and also develops and provides centrally audit analytics. There is a high degree of specialization within the team. In addition to a Shared Service Center, the Audit Intelligence Team is also professionally supported by Bayer Business Service GmbH, an internal IT service provider. The IT backbone and application are centrally hosted and supported in this service company.

Data Privacy Compliance with the General Data Protection Regulation (GDPR) of the European Union, which had to be implemented until May 25, 2018, must be ensured. This regulation must be adhered to from the outset during conception and design and in cases of any doubt the data privacy responsible of Bayer AG has to be contacted.

Release Change As mentioned before, Internal Audit's Process Mining application is not hosted in a cloud, but on premise. That means that release changes have to be executed manually, inclusive of testing, which has turned out to be a time- and effort-consuming activity.

Outlook

Process Mining is a groundbreaking approach for Internal Audit, as the event-log-based process representation opens up completely new perspectives for the representation and auditing of processes. Especially for global players and their bundling of processes in offshore shared service centers and regional platforms, Process Mining provides the basis for establishing a more effective process audit. If the legal entity is in most cases the subject of prevailing audits, Process Mining opens up completely new perspectives for the design of audits: the audit of the global or regional cross legal entity process. The fact that the MS SQL Server hosts business data of 250 legal entities allows completely new possibilities of auditing: Many audits could be carried out centrally in the form of desktop audits at the Bayer HQ in Leverkusen, so that audits in countries or legal entities are only required if specific questions (qualitative matters, i.e., deal making) arise that cannot be clarified on the basis of SAP data. Centrally executed audits are conceivable as an interplay between data mining and Process Mining. In the end, the key question that must be answered is what should be audited centrally and which residual topics should be audited locally in the legal entities/the countries.

As far as the follow-up of an audit is concerned, Process Mining also offers extended possibilities, in particular in the quantifiable area of action items. For example, the question of whether goods receipts are carried out according to recommendation xyz can be answered via Process Mining. In this way, the implementation of a recommendation or action item can be checked without time-consuming communication with the auditee.

As far as the perspectives of Process Mining are concerned, it should be continuously determined to what extent Process Mining is scalable, i.e., which processes can be represented with the help of event logs. In addition to scalability, the question should also be raised to what extent Process Mining can be combined with other innovative technologies. In addition to buzz words such as AI or machine learning, a focus on the topic of RPA seems promising. By means of Process Mining, the As-Is process flows can be represented in a first step. In a second step process deviations can be identified—supported by conformance analyses. Those audit steps that have a high frequency and which have so far been performed manually but show a high potential to be standardized, could be automated with the help of RPA and thus lead to a relief of the auditors. This idea is reflected in PWC's statement: "And, perhaps the greatest opportunity: testing of controls and other departmental tasks can be automated through RPA, expanding internal audit's capacity and freeing auditors to focus on more value-added activity" (PWC, 2017). As indicated in the technology part, RPA is also conceivable for the purpose of data validation.

Finally, the benefits of Process Mining are to be considered from a broader perspective. Process Mining contributes to the visualization and identification of (process-related) risks. As valuable as the application may be—the application of Process Mining related to Internal Audit basically ends here. However, Process Mining is much more than a discovery software that enables the identification of process deviations and risks. After the identification of deviations from the To-Be process, the process enhancement phase follows, in which the previously identified improvement potentials are tapped. As Internal Audit is not responsible for processes in the narrower sense, there is only the possibility to positively influence process quality, process compliance, and risk minimization by means of recommendations and action items. Although Process Mining can only be used to its full extent in business or by (Global) Process Managers, it represents an enormous additional benefit for Internal Audit functions.

Links

Fleischmann, A. et al. (2018): Ganzheitliche Digitalisierung von Prozessen, Wiesbaden.

Jans, M./Hosseinpour, M. (2019): How active learning and process mining can act as Continuous Auditing catalyst, in: International Journal of Accounting Information Systems, No 32 (2019), pp. 44–58.

Maldonado-Mahauad, J. et al. (2017): Mining theory-based patterns from Big data: Identifying self-regulated learning strategies in Massive Open Online Courses, 2017 (accepted manuscript).

Chiu, T. (2019): Performing Tests of Internal Controls Using Process Mining, in: The CPA Journal, June 2019, pp. 54–57.

van der Aalst, W. (2016): Process Mining, Heidelberg.

van der Aalst, W. (2012): Towards improving the representational bias of process mining, in: Data-driven Process Discovery and Analysis, Springer, pp. 39–54.

URLs

Delias, P. (2018): Visualizing and exploring event databases: a methodology to benefit from process analytics, in: Operational Research, URL: https://doi.org/10.1007/s12351-018-00447-z, as of 2019-11-24.

PWC (2017): Robotic process automation: A primer for internal audit professionals, URL: https://www.pwc.com/us/en/risk-assurance/publications/assets/pwc-robotics-process-automation-a-primer-for-internal-audit-professionals-october-2017.pdf, as of 2019-11-20.

Jans, Mieke (2015): https://www.youtube.com/watch?v=dwel6OEJ-0A, Mieke Jans—Process Mining Camp 2015, 52 min 14 sec, as of 2019-11-17.

Telekom: Process Mining in Shared Services 20

Gerrit Lillig

Abstract

In 2016, Deutsche Telekom Services Europe decided to improve the analytics capabilities in one of the most important internal e2e processes. As a shared service center is typically focused on e2e process performance, one major attempt was the implementation of a Process Mining software in order to further improve the efficiency. The idea was to investigate our core processes, to find out where to shorten lead times, reduce complexity, and make the processes more efficient. During the implementation it then turned out that our shared service could benefit far more from this technology: We built operational steering capabilities, which led to concrete savings. We were able to bring our reporting and analytics capabilities on a new level. And we helped to position our shared services internally as a driver for digitalization. Of course, the road towards this was paved with a lot of challenges like workers' council negotiations, internal constraints, and technical challenges—just to mention a few of them. At the end that all paid-off—we saved a lot of money in our operations, were able to establish a new digital steering solution, and we now have a flexible and powerful reporting solution at hand.

Challenge

The beginning of our Process Mining journey at Deutsche Telekom was in 2016. Deutsche Telekom had decided to build a new Multi-Shared Service Center, covering procurement, accounting, human resource, and reporting services for our internal partners and customers. This service was supposed to be offered by Deutsche Telekom Services Europe (DTSE), a new internal organization with a new service

G. Lillig (✉)
Telekom Deutschland GmbH, Frechen, Germany
e-mail: Gerrit.lillig@telekom.de

© Springer Nature Switzerland AG 2020
L. Reinkemeyer (ed.), *Process Mining in Action*,
https://doi.org/10.1007/978-3-030-40172-6_20

delivery, organized in eight e2e-oriented service lines. The migration and transformation towards this new target organization (which came along with reorganization, site and building changes, changes in responsibility, nearshoring, etc.) was planned to take approximately 3 years and came along with a certain efficiency and quality ambition. At that time, I had the opportunity to take over the department, which was responsible for operational steering and analytics in the service line Procure-to-Pay (PTP). PTP was the new flagship service line of DTSE, as it was the first real e2e-oriented service line and was put together out of several departments and units from different sources. Accordingly, this new service line was faced with the challenge of developing a new e2e-oriented steering concept. So far there had been only functional reporting solutions in procurement, accounts payable, banking, and support services in place. Most of them were only showing a selected subset of functional KPIs and even many of them were not up to date due to a huge backlog of query development on the IT side. It turned out that due to prioritization and resource shortage on the IT side in many cases the maintenance of KPIs and queries had been deprioritized against functional requirements in core-systems and thus only a limited number of BI-developers were available. Even worse was the situation in our process management from a process documentation perspective. The available process models in Procurement and Accounts Payable didn't fit together at all—one was rather functional (developed as a basis for IT implementation), whereas the other was supposed to guide the customers through the order management process. What both had in common: they were outdated and not up to date at all.

With one's back to the wall we decided to try the forward escape—looking for an e2e-based process analytics solution, which allowed us to optimize, report, and steer our processes and could be maintained by the business department, without too much support from IT developers. Content-wise it was important not only to replace the functional KPIs (e.g., invoice automation) but to take a look at the e2e process (e.g., how automated is PTP overall and how is the perceived lead-time from customer perspective?). Accordingly, we shifted the focus from a functional to a process-oriented perspective. In addition to that we decided to also change the approach on how we try to optimize our processes. We got rid of the idea to maintain process models in Business Process Model and Notation (BPMN) which would try to optimize our processes based on our understanding how the processes should run. Instead, the new approach should focus on how the processes really run in the system—based on the digital footprints of the processes in our system landscape.

At that point of time we already had a pilot running with a Process Mining vendor in one of our cross-functional departments and have had an intensive exchange with the software provider of our process documentation solution. In fact, at that time both solutions seemed to be quite complex and difficult to be set up and maintained. As a third try we decided to pilot our PTP process with a German software provider, who offered a good SAP integration and a high user experience. On top of that they had a proven use case with one of our biggest Telco competitors on the German market. If it works with the main competitor—why shouldn't it work at our company? Even more the most attractive point for us was the promise of intuitive tool, which could be implemented rather without expert know-how and too much IT effort (this was by the way the reason the implementation project was called PROMISE—Process Mining Innovation at Services Europe).

The journey we had to undertake in order to get the pilot done was incredibly long. Due to internal regulation on data privacy, data security, and alignment with the workers' council it took us half a year to get all the necessary approvals to run a pilot on a sample set based on real data. From time to time we discussed internally to change the approach and to run the pilot with synthetic data. But in the end our approach turned out to be the right one! The results of the pilot were amazing: even on the pilot data we could directly see possible process improvements and potential to steer our processes more efficiently.

By the way: after we got all the approvals, it took the project team 1 week to provide the final process analyses on our sample data! As the results were very promising, we decided to go directly for a complete implementation. I will never forget the blink in the eyes of my reporting colleagues, who had been working on BI solutions for decades, when they saw the first results of the process analyses and the incredible speed of the tool.

Use Case

The general promise of a Process Mining solution is to help your company in optimizing process efficiency. For our PTP processes, I have to admit that this was not the real trigger for us. As we had already gone through several efficiency programs, our PTP processes were at that time already on an efficiency level, which was benchmarked by The Hackett Group at the top level in our peer group. In contradiction to the classical purpose of Process Mining, we rather aimed at improvement of process steering based on analytics. Especially in the PTP process our process efficiency and process effectiveness was already benchmarked to be far ahead the peer levels. Nevertheless, the effects we planned to achieve out of an optimized steering approach already showed to turn the business case of the implementation into a positive RoI.

Accordingly, we started with a Process Mining implementation on the PTP process. To keep the implementation manageable, we started with the implementation of our core-system, which covers the majority of the PTP processes of the DT group. From the early beginning one major goal of the project was to build up internal competence in order to be able to set-up further processes without further support. Therefore, we decided to run the project with as less external support as possible. On top of that all external activities needed to be focused on knowledge transfer towards internal colleagues. Today, 3 years after this decision I can indeed say that this decision was one of the best decisions we took during the project.

Our implementation roadmap within the group can from today's perspective be clustered in four steps.

Step 1: The First Implementations

As soon as it turned out that our PTP implementation was about to be successful, we already started the following process implementations: Buy-to-Scrap (Plan, Build, and Run the network) was the next candidate, followed by further PTP processes from further systems and (due to extensive discussions with the workers' council) 2 years later also a couple of HR processes. Step by step we applied our Process Mining solution to nearly all relevant delivery processes which our Shared Service Center is in charge of.

Step 2: Center of Excellence

During the implementation we had various training sessions with internal customers of our Shared Service Center, who were then not only interested in the results of Process Mining on our shared service center processes. There was also an interest to try out the software on core and support processes of our national and international segments. This in fact turned out to be an interesting chance for our Shared Service Center. With the major customers we agreed to set up a central Center of Excellence (CoE) for Process Mining, which got the task to implement Process Mining group wide. This was at that time not only an additional service for our shared services but even a more fantastic opportunity to position DTSE as a driver for digitalization within the group.

Within half a year we took the necessary steps to build a new CoE for Process Mining. As we knew that the HR market for data engineers and data experts in Germany was rather difficult, we directly decided to go for a two-location approach. After a short screening by our HR experts it turned out that our site in Brno in Czech Republic seemed to be a good location to acquire young talents. In fact, it took us roughly 1 year to get our CoE running in a stable mode—especially the onboarding of new colleagues, establishment of procedures and structure, and the international collaboration turned out to be the major challenges.

The CoE provides support in three major areas. First it is able to steer implementation projects together with the affected Business team and the IT experts, which we need for data provisioning. Nevertheless, even though the knowledge of the project methodology to implement Process Mining is allocated in the CoE, it is important that the project lead is assigned to the business team. This ensures the necessary buy-in and drive at the respective business area. The second offering of the CoE is central data engineering and data science expertise. Together with the business analysts from the business teams, the CoE is able to set up the data models and event log procedures in our Process Mining solution. Finally the CoE also offers first- and second-level support for the tool. We integrated this into the existing support-landscape of our shared service center, so that support requests regarding our Process Mining software directly go into our salesforce service cloud and can be solved there by dedicated and trained support agents (Fig. 20.1).

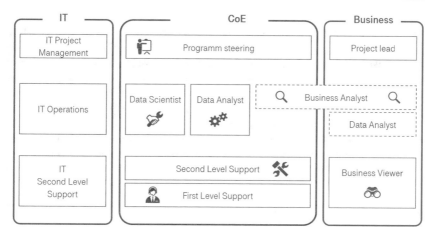

Fig. 20.1 Center of Excellence for Process Mining

The long-term results of that decision were from my perspective fantastic. Firstly it allowed us to build internal competence to build new process models in-house. Secondly (beside two to three parallel innovative projects) it had a huge impact to reshape the perception of our internal customers and stakeholders towards our shared service center. We were not so much considered to be the "cost case" and the "efficiency company" anymore but were a well-recognized partner in terms of innovation and optimization. And this finally helped us in our recruiting. The applications (not only in the area of data analytics) we had at that time were of a much higher quality than in the time before. It started to become attractive to work at DTSE.

Step 3: Integration

The third step was then mainly characterized by integration. First, we integrated our Process Mining solution with our process alerting engine, which we had implemented in parallel. The process alerting engine was supposed to create alerts, based on preconfigured business rules. For example, we configured a process alert in order to prevent the circumvention of approval thresholds in the PTP chain. In fact, whenever there are several similar shopping carts created with comparable content, ordered by the same party, an alert is created and going to be checked by a process agent. After we had integrated this solution with our Process Mining solution, we could not only show our alert statistics in the Process Mining tool, we also had the possibilities to do process analytics on the group of processes, for which we had seen similar alerts. In terms of compliance improvement, this is a fascinating opportunity.

The second aspect of integration referred to the implementation of Process Mining on integrated process models. The idea behind that use case is to interlink interdependent process chains. So far Process Mining was usually focused on classical transactional chains, which followed a certain e2e process path, e.g., PTP, O2C, or Customer Service. Also, our Customer Journey started with PTP and at some time also with a Customer Service Use Case for our PTP support processes. As soon as we had this Customer Service use case live, we started to think about how we could find out the real root causes for process delays in the PTP chain and which factors might influence customer satisfaction. This was the birth-hour for our idea to go for a horizontal and vertical integration in process analytics (the delivery process here is considered to be horizontal, crossed by the vertical support process). A practical example is an analysis that shows which combination of process patterns (e.g., certain material groups at a certain time from a concrete vendor) lead to complaints in the invoice handling process.

Step 4: Machine Learning

The last step then was even the most exciting one. As one of the first pioneers (at least for ML use cases with our software provider) we started to realize first predictive analytics and machine learning use cases. The idea behind that and the use cases we had in mind were fascinating. But to be honest: The time until we had the infrastructure available was incredibly long and the way to get it finally running was sometimes quite frustrating. I don't remember anymore how often we had to re-setup and install machine learning servers and fix configuration issues. But in the end (and especially due to of a couple of selected colleagues) it worked, and we had a Process Mining software with the capability to use R and Python algorithms. The positive aspect of this delay was that we had enough time to get the necessary approvals from workers council, data security, and data privacy, so that we were able to start at the point of time when the technical solution was available. Today—about 1 year later— we have the first productive use cases running on that environment, which proves to be a game changer in our financial steering. For example, we have built a process steering engine for our capital expenditure controlling, which predicts the payment date and amount of CAPEX-relevant invoices (including a Machine Learning algorithm, which detects whether an invoice will be CAPEX or OPEX and certain predictors regarding invoicing behavior of vendors) as shown in Fig. 20.2.

Impact

Besides the already mentioned achievements, we indeed experienced a measurable improvement in process quality by steering the process with our Process Mining tool. As one example: Our cash discount rate could be increased remarkably, due to the fact that our team responsible for checking overdue invoices, monitored invoices at risk on a daily basis. Besides the better quality, the improved steering and a couple of optimization measures led to operational savings of much more than one million euros.

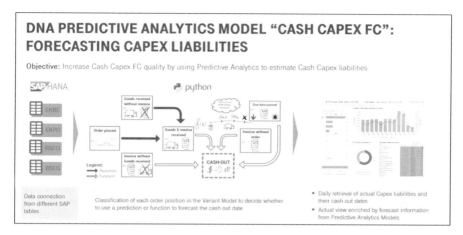

Fig. 20.2 Cash CAPEX use case

On top of that we experienced a massive increase in customer satisfaction. Whereas in former times our customers had no transparency in how their processes executed by DTSE were run, we were now able to offer 100% transparency (and in fact we did that and provided them access to the process monitor). Apart from that due to the new setup and the new roles and expertise we gained in our shared service, there was hardly any reporting request anymore which I couldn't answer within 1 day.

And last but not least our effort invested in stage four was the door opener for advanced analytics at DTSE: Since then we have been able to realize a couple of use cases due to the fact that we have the infrastructure at hand, we have colleagues with the necessary expertise, and an environment setup, where the permissions from data security, data privacy, and workers council are available.

Technology

As we were focused on our shared service center processes in the beginning, we took the decision to integrate our Process Mining engine into our shared service BI system landscape. This decision at that time followed the recommendation of our IT architects and by the software vendor. It was mainly based on the reason to ease the implementation with our IT and to ease also the negotiation with the workers' council. Apart from that even our first use case required to analyze a process which went through three different systems—therefore a consolidating layer between the transactional systems and the Process Mining engine was a must.

Today our architecture looks as shown in Figs. 20.3 and 20.4: The transactional systems (mainly SAP Systems, but also other systems like Salesforce) are connected

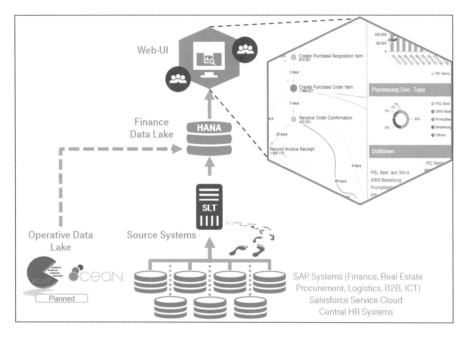

Fig. 20.3 Telekom architecture

to a central SAP HANA Data base. Our Process Mining software is installed onsite and connected to that HANA database for loading the data. On that HANA Database we have also realized our process alerting engine. The alerts are so far sent to a SAP GRC solution for alert handling, which will be replaced by our Salesforce Service Cloud quite soon.

For predictive analytics and machine learning purposes, we have meanwhile also a Machine Learning Server installed on our BI landscape to make use of R and Python algorithms. The next step should be to connect our Hadoop Cluster to this landscape, as many customer-oriented processes are running on that data lake as well. We are currently working on that as well, but still must tackle a couple of regulatory and technical issues.

Lessons Learned

After having Process Mining in use for 3 years now, we have collected a couple of dos and don'ts. One main recommendation for companies who are going for an implementation now would be to think very early how the architecture should look like. Cloud solutions offered today seem to be very interesting in terms of operation and offered functions. I do nevertheless not think that we would have been able to

Fig. 20.4 Alert engine

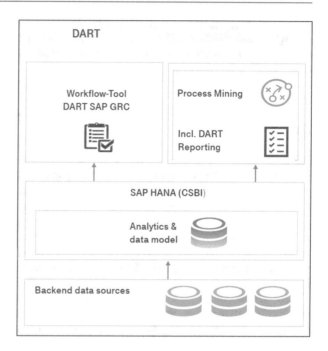

realize so many use cases—even on HR and Service processes in the cloud—without getting into massive troubles with our data security and data privacy constraints. Therefore, it is important to think early about how use cases shall look like, how important it is to work with real (or maybe at least anonymized data), and how far this fits into a cloud solution.

One of our most important success factors was a very early involvement of the key stakeholders. First: the workers' council. We offered our workers council to become part of the project which was at that time a bit disruptive. Instead of developing a concept and ask for approval, we made some workers council colleagues an integral part of the team. This helped us to identify early which parts of our tool were not acceptable (and of course we took them out)—on the other hand, this made the negotiation and approval much easier, as the colleagues had a realistic picture of what the tool was capable of and that it won't be used for performance control and monitoring of employees.

The second success factor was to think directly from the early beginning from an operations perspective. How shall the tool and the data models be used and maintained in daily life after implementation? One major attempt was to accordingly build data engineering and data science expertise from the early beginning during the implementation. This helped us learn during the project phase and to implement new use cases in short time frames without external support later.

And finally, I would recommend having a realistic view and expectation management of what the tool is capable of. It is important to considered that Process

Mining is not plug and play. It needs certain time to set up process models and to configure and validate them. Therefore, our approach typically is to implement rather with a small effective team. One colleague from the affected business department needs to be part of the team, who needs the use case later. This colleague can act as an influencer and get the acceptance of the involved departments later.

Outlook

After we have implemented Process Mining on a broad basis and a variety of processes (from PTP, Customer Service, HR, and O2C to Real Estate Management), the next level will be further connection of dependent use cases. This follows the idea to run Process Mining on processes which do not follow an e2e flow in the system. A concrete idea is, e.g., to follow the customer journey process. The process might start with a first touch point in a T-shop (or with a service agent) where a mobile phone with contract is being ordered, goes on via logistic delivery, and ends with the first invoice being paid. So far this would mean for us three separate Process Mining models—in the future we would like to see it in one.

Our second focus is to further develop our Process Mining solution with the capabilities of machine learning. First, we plan to automate pattern recognition within the Process Mining tools. Combinations of triggers and root causes for inefficiencies as well as correlations between them should in the future be determined by the tool itself. In the medium term we would use the power of Process Mining and machine learning for prescriptive analytics, which allows us to focus on the crucial items to steer the process automatically.

Links

Interview at the SAPHIRE Now conference 2018: https://www.youtube.com/watch?v=wMGk8TZFG10&feature=youtu.be
Process Mining Story Telekom: Turning Process Transparency into Procurement Leadership: https://www.youtube.com/watch?v=k7ldNIRz99A
LinkedIn Profile: https://www.linkedin.com/in/gerrit-lillig/
Article from our CFO of Telekom Deutschland: https://www.linkedin.com/pulse/process-mining-organic-transformation-opportunity-klaus-werner
Interview on Digitalization (in German language only): https://beschaffung-aktuell.industrie.de/news/digitalisierung-ist-kein-selbstzweck/

Part III

Outlook: Future of Process Mining

Academic View: Development of the Process Mining Discipline

21

Wil van der Aalst

Abstract

This chapter reflects on the adoption of traditional Process Mining techniques and the expansion of scope, discussed with five trends. An inconvenient truth explains why—despite considerable progress in Process Mining research—commercial tools tend to not use the state-of-the-art and make "short-cuts" instead that seem harmless at first, but inevitably lead to problems at a later stage. Seven novel challenges provide an outlook on open research topics. In a final appeal, the term of "process hygiene" is coined to make Process Mining the "new normal."

It is exciting to see the spectacular developments in Process Mining since I started to work on this in the late 1990s. Many of the techniques we developed 15–20 years ago have become standard functionality in today's Process Mining tools. Therefore, it is good to view current and future developments in this historical context.

This chapter starts with a brief summary of the history of Process Mining, showing how ideas from academia got adopted in commercial tools. This provides the basis to talk about the expanding scope of Process Mining, both in terms of applications and in terms of functionalities supported. Despite the rapid development of the Process Mining discipline, there are still several challenges. Some of these challenges are new, but there are also several challenges that have been around for a while and still need to be addressed urgently. This requires the concerted action of Process Mining users, technology providers, and scientists.

W. van der Aalst (✉)
RWTH Aachen University, Aachen, Germany
e-mail: wvdaalst@pads.rwth-aachen.de

© Springer Nature Switzerland AG 2020
L. Reinkemeyer (ed.), *Process Mining in Action*,
https://doi.org/10.1007/978-3-030-40172-6_21

Adoption of Traditional Process Mining Techniques

Process Mining started in the late 1990s when I had a sabbatical and was working for 1 year at the University of Colorado in Boulder (USA). Before, I was mostly focusing on concurrency theory, discrete event simulation, and workflow management. We had built our own simulation engines (e.g., ExSpect) and workflow management systems. Although our research was well received and influential, I was disappointed by the average quality of process models and the impact process models had on reality. In both simulation studies and workflow implementations, the real processes often turned out to be very different from what was modeled by the people involved. As a result, workflow and simulation projects often failed. Therefore, I decided to focus on the analysis of processes through event data [1]. Around the turn of the century, we developed the first process discovery algorithms [2]. The Alpha algorithm was the first algorithm able to learn concurrent process models from event data and still provide formal guarantees. However, at the time, little event data were available and the assumptions made by the first algorithms were unrealistic. People working on data mining and machine learning were (and perhaps still are) not interested in process analysis. Therefore, it was not easy to convince other researchers to work on this. Nevertheless, for me, it was crystal clear that Process Mining would become a crucial ingredient of any process management or process improvement initiative.

In the period that followed, I stopped working on the traditional business process management topics and fully focused on Process Mining. It is interesting to see that concepts such as conformance checking, organizational Process Mining, decision mining, token animation, time prediction, etc., were already developed and implemented 15 years ago [2]. These capabilities are still considered to be cutting-edge and not supported by most of the commercial Process Mining tools.

Figure 21.1 illustrates the development of the field. On the one hand, the graph shows the growth of the scientific Process Mining literature. Each year, a growing number of Process Mining papers are published in journals and presented at

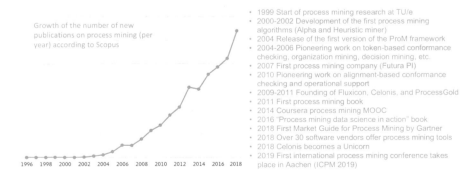

Fig. 21.1 Summary of the history of Process Mining, also showing the growth of scientific papers on the topic

conferences. On the other hand, the right-hand side of the figure also mentions a few milestones illustrating the uptake in industry. The first Process Mining company (Futura PI) was founded in 2007 by one of my students (Peter van de Brand). The software was later integrated into the tools of Pallas Athena and Perceptive Software. A few years later, Fluxicon, Celonis, and ProcessGold were founded. Concurrently, the first Process Mining books appeared and the first online course for Process Mining was created (followed by over 120,000 participants). However, until 2015, the practical adoption of Process Mining in industry was limited. Only in recent years the actual usage accelerated [3, 4]. This is illustrated by the growing number of Process Mining vendors. Currently, there are over 30 Process Mining vendors (e.g., Celonis, Disco, ProcessGold, myInvenio, PAFnow, Minit, QPR, Mehrwerk, Puzzledata, LanaLabs, StereoLogic, Everflow, TimelinePI, Signavio, and Logpickr). In 2018, the first International Conference on Process Mining (ICPM) was organized, illustrating the growing maturity of the field. Moreover, as this book shows, there are many exciting applications in organizations such as Siemens, BMW, Uber, Springer, ABB, Bosch, Bayer, Telekom, etc.

Although some Process Mining vendors have added conformance checking techniques and more advanced discovery techniques like the inductive mining approach, the basis of most commercial Process Mining tools is still the *Directly-Follows Graph* (DFG). This was actually the graph that served as input for the classical Alpha algorithm 20 years ago. The DFG can also be viewed as a traditional transition system or a Markov chain (when adding probabilities). A DFG is a graph with nodes that correspond to activities and directed edges that correspond to directly-follows relationships [2, 5]. The frequency on an arc connecting activity X to activity Y shows how often X is directly followed by Y for a specific case (i.e., process instance). Similarly, the arc can be annotated with time information to show bottlenecks. Using frequencies it is possible to seamlessly simplify such process models. It is also possible to animate the cases using tokens moving along the directed arcs. This is easy to understand and highly scalable. Therefore, this basic functionality is present in all of today's Process Mining tools.

Expanding the Scope of Process Mining

Over the last two decades, the scope of Process Mining expanded in different ways. First of all, Process Mining grew out of academia into industry. Also the number of application domains expanded [6]. Traditionally, applications were limited to financial or administrative processes. The Order-to-Cash (O2C) and Purchase-to-Pay (P2P) processes are obvious candidates to apply Process Mining. However, nowadays Process Mining is also applied in healthcare, logistics, production, customs, transportation, user-interface design, security, trading, energy systems, smart homes, airports, etc. This makes perfect sense since event data and processes are not limited to specific application domains. Finally, there is also a clear expansion in the capabilities of Process Mining tools. Initially, the focus was exclusively on process discovery based on historic data. However, the scope of Process Mining expanded

Fig. 21.2 Four trends showing that the scope of Process Mining is expanding

far beyond this as is illustrated by the four trends depicted in Fig. 21.2 and discussed next.

The scope of Process Mining expanded from a tool for a data scientist used in an improvement project to the *enterprise-wide, continuous* application of Process Mining. Process Mining tools only supporting the construction of DFGs with frequencies and times tend to be used in smaller projects only. These projects often have a limited scope and duration. As a result, few people use the results and there is no support for continuous improvement. Given the investments needed for data preparation, it is often better to apply Process Mining at an enterprise-wide scale with many people using the results on a daily basis. It does not make sense to see Process Mining as a one-time activity. However, the enterprise-wide, continuous application of Process Mining requires substantial resources in terms of computing power and data management (e.g., to extract the data and convert these into the right format). Moreover, to lower the threshold for a larger Process Mining community within an organization, one needs to create customized dashboards. Therefore, Process Mining needs to be supported by higher-level management to realize the scale at which it is most effective.

Initially, Process Mining efforts focused on process discovery [2]. However, over time it has become clear that *process discovery is just the starting point* to process improvement. One can witness an uptake in conformance checking and performance analysis techniques. Moreover, Process Mining is often combined with data mining and machine learning techniques to find root causes for inefficiencies and deviations. Although process discovery will remain important, attention is shifting to the steps following discovery using optimization, machine learning, and simulation.

A third trend is the shift in focus *from backward looking to forward looking*. Traditional Process Mining techniques start from historic data. This can be used to diagnose conformance and compliance problems. However, organizations are often more interested in what is happening now or what is going to happen next. Backward-looking Process Mining can be used to fundamentally improve processes, but provides little support for the day-to-day management of processes. Therefore, event data need to be updated continuously and Process Mining techniques need to be able to analyze cases that are still running. This is needed to control and influence the running process instances. Some of the commercial Process Mining tools provide excellent capabilities to show the current state of the process. A next step is the application of more forward-looking techniques able to predict what will happen to individual cases and where bottlenecks are likely to develop. Techniques for operational support (i.e., detecting compliance and performance problems at runtime, predicting such problems, and recommending actions) have been around for more than a decade. However, their quality still leaves much to be desired. The problem is that cases (i.e., process instances) highly influence each other when competing for resources. Moreover, there may be a range of contextual factors influencing processes. By simply applying existing machine learning and data mining techniques, one cannot get any reliable predictions. Hence, additional work is needed.

The fourth trend is the increased focus on actually improving the process. Process Mining tends to focus on diagnostics and not on the *interventions* that should follow. Insights generated by Process Mining should be actionable. Therefore, Process Mining is increasingly combined with *Robotic Process Automation (RPA)*. Process Mining can be used to identify manual tasks that can be automated and monitor software robots. This allows for the automation of processes for which traditional workflow automation would be too expensive. Figure 21.3 shows the spectrum of process variants. High-frequent variants are candidates for automation, but lower frequent variants cannot be automated in a cost-effective manner. RPA helps to shift the boundary where (partial) automation is still cost-effective. Interestingly, Process Mining can be used before and after automation for any mixture of process variants (automated or not). However, RPA is just one of many ways to turn Process Mining diagnostics into actions. Here, I would also like to coin the term *Robotic Process Management (RPM)* to refer to automatic process interventions. Unlike RPA, RPM is not automating steps in the operational process. RPM translates Process Mining diagnostics into management actions. For example, when a bottleneck emerges RPM may take actions such as alerting the manager, informing the affected customers,

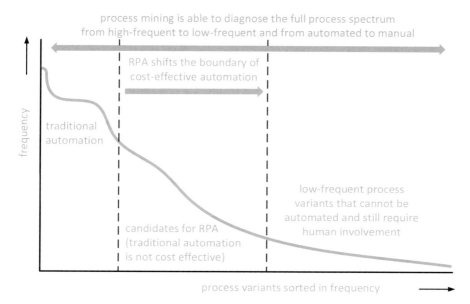

Fig. 21.3 Process Mining can be used to identify candidates for RPA and monitor any mixture of automated/non-automated frequent/infrequent process variants

assigning more workers, etc. Another example would be to prioritize targeted auditing activities when a significant increase in process deviations occurs. These examples show that the diagnostics provided by Process Mining are just the starting point.

An Inconvenient Truth

Despite the rapid developments in Process Mining, many of the original challenges remain [2]. Although there has been considerable progress in Process Mining research, commercial tools tend to not use the state-of-the-art due to pragmatic reasons such as speed and simplicity. Commercial software tends to make "short-cuts" that seem harmless at first, but inevitably lead to problems at a later stage.

The first inconvenient truth is that *filtered Directly-Follows Graphs (DFGs) have well-known problems*. DFGs cannot express concurrency and filtering them may provide misleading results. Yet, the default discovery techniques used by commercial tools are all based on filtered DFGs.

To illustrate the problem, consider a fictive purchasing process consisting of five steps: *place order*, *receive invoice*, *receive goods*, *pay order*, and *close*. In this idealized process, all five activities are performed for all procurement orders. The process always starts with activity *place order* and ends with activity *close*. However, the three middle activities are executed in any order. For example, in rare cases the organization pays before receiving the invoice and goods. Hence, there are six different process variants. In our event log there is information about 2000 orders, i.e., in total there are 10,000 events. The most frequent variant is ⟨*place order*, *receive invoice*, *receive goods*, *pay order*, *close*⟩ which occurs 857 times. The least frequent variant is ⟨*place order*, *pay order*, *receive goods*, *receive invoice*, *close*⟩ which occurs only four times.

Figure 21.4 shows the process model discovered by ProM's Inductive Miner [2]. The three unordered activities in the middle are preceded by an AND-split and are followed by an AND-join. The model correctly shows that all activities happen precisely 2000 times. Applying other basic process discovery algorithms like the Alpha algorithm and region-based techniques yield the same compact and correct process model (but then often directly expressed in terms of Petri nets).

Let us now look at the corresponding Directly-Follows Graphs (DFG) used by most commercial Process Mining tools [2]. Figure 21.5 shows three DFGs generated

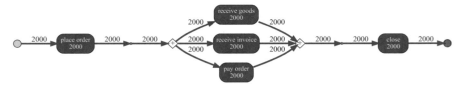

Fig. 21.4 Process model discovered by ProM's Inductive Miner. Note that all activities occur once per procurement order

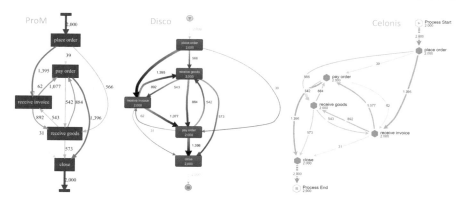

Fig. 21.5 Three identical DFGs returned by ProM, Disco, and Celonis

by ProM, Disco, and Celonis. Apart from layout differences, these three models are identical. Surprisingly, the DFGs suggest that there are loops in the process although each activity is executed precisely once for each order. The process also seems more complex. To simplify such DFGs, all Process Mining tools can leave out infrequent paths to simplify the model. However, this may lead to highly misleading results, e.g., frequencies no longer add up and averages are based on unclear fragments of behavior.

To further illustrate the problem, we now consider a variant of the order process where each of the middle activities is skipped with 50% probability. This means that for approximately 50%·50%·50% = 12.5% of cases only the activities *place order* and *close* are performed. Figure 21.6 shows the process model discovered by ProM's Inductive Miner clearly showing that the three middle activities can be skipped. For example, the process model shows that for 988 of the 2000 orders there was no payment. Figure 21.7 shows the corresponding Petri net model without frequencies.

As before, we can also discover the DFG for this second event log. The result is shown in Fig. 21.8. Again ProM, Disco, and Celonis generate identical DFGs. We can see that for 236 orders, all three middle activities are skipped. Again we see

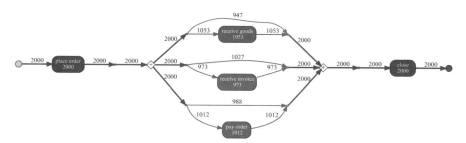

Fig. 21.6 Process model discovered by ProM's Inductive Miner for the event logs where the middle three activities can be skipped

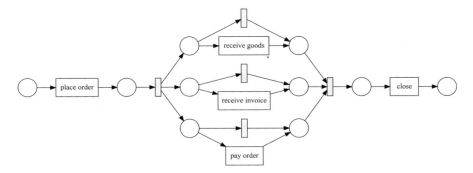

Fig. 21.7 Petri net discovered by ProM showing that each of the three middle activities can be skipped

loops that do not exist and the underlying structure of the process clearly depicted in Figs. 21.6 and 21.7 is invisible in the three DFGs.

As mentioned, DFGs can be seamlessly simplified by leaving out infrequent activities and arcs. Of the three middle activities, activity *receive goods* is most frequent. Hence, this activity remains when we filter the model to retain the three most frequent activities. Activities *place order* and *close* occur 2000 times, and activity *receive goods* occurs 1053 times. Figure 21.9 shows the filtered DFGs generated by Disco and Celonis. Now there are some surprising differences. These differences illustrate that filtered DFGs can be very misleading.

Note that Figs. 21.8 and 21.9 are based on the same event data and the frequencies of activities are the same in all DFGs shown, e.g., activity *receive goods* occurs 1053 times in all DFGs. However, the information on the arcs is very different. Let us first focus on the connection between activity *place order* and activity *close*. Activity *place order* directly followed activity *close* in 236 cases. This was correctly shown in Fig. 21.8. However, Celonis reports the same number (i.e., 236) when the other activities have been removed (Fig. 21.9). However, this can of course no longer be

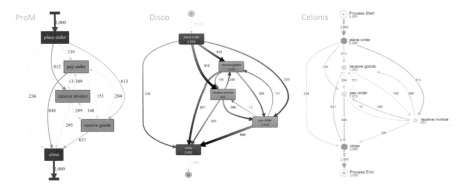

Fig. 21.8 Three identical DFGs for the process where the three middle activities can be skipped

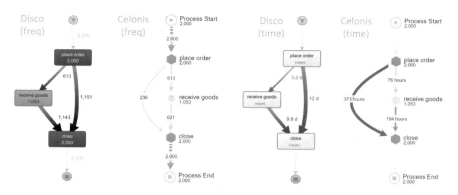

Fig. 21.9 Filtered DFGs generated by Disco and Celonis showing frequencies (left) and time (right)

the case. After removing *receive invoice* and *pay order* there are more cases where *place order* is directly followed by activity *close*. Also, Disco reports an incorrect number (i.e., 1151). After removing the two activities, there are 947 cases where *place order* is directly followed by activity *close* and the average time between these activities for these cases is 13.8 days. Surprisingly, Disco reports 12 days and Celonis reports 15.5 days. Hence, Celonis and Disco report different frequencies and times and both fail to show correct values for frequencies and times.

Another example is the connection between activity *receive goods* and activity *close*. If we project the log onto the three remaining activities, we can see that there are 1053 cases where activity *receive goods* directly follows activity *close* and this takes on average 8.5 days. Disco reports a frequency of 1143 (too high) and an average time of 9.8 days (too high). Celonis reports a frequency of 621 (too low) and an average time of 8 days (too low).

Moreover, the diagrams in Fig. 21.9 are internally inconsistent. The frequencies involved in split and join should add up to 2000, but we can witness 613+1151 and 236+613 for the split and 1143+1151 and 236+621 for the join. These inconsistencies will confuse any user that would really like to understand the process. Figure 21.10 shows that this is not necessary.

These small examples illustrate an inconvenient truth. Simplistic discovery techniques providing just a DFG with sliders to simplify matters are not adequately capturing the underlying process. The frequencies and times reported are wrong (or at best misleading) and when activities are not executed in a fixed order there will

Fig. 21.10 Process model discovered by ProM's Inductive Miner showing the correct frequencies after abstracting from the two lower frequent activities

always be loops in the model even when these do not exist in reality. These problems are not specific for Disco or Celonis. Almost all commercial Process Mining tools make shortcuts to ensure good performance and provide similar results. Although this is a known problem that has been reported repeatedly over the last decade, vendors are reluctant to address it. There are two main reasons: *simplicity* and *performance*. Petri nets, BPMN models, UML diagrams, etc., are considered to be too complicated for the user. However, the price to pay for this simplicity is the presence of spaghetti-like diagrams with many nonexisting loops. To ensure good performance, filtering of the DFG is done on the graph rather than on the original data. This explains the incorrect frequencies and times in Fig. 21.9.

Another inconvenient truth is the limited support for *conformance checking* [2]. Although conformance checking is considered to be important from a practical point of view, it is still not very well supported and rarely used. Conformance checking approaches ranging from token-based relay [7] to alignments [8] have been developed over the past two decades. Several vendors try to support conformance checking by comparing discovered DFGs with normative DFGs derived from hand-made process models. This, of course, does not work. Compare, e.g., the DFGs in Fig. 21.5 with the DFGs in Fig. 21.8. The only difference is the connection between activity *place order* and activity *close*. However, it is very difficult to see that in the second data sets all possible subsets of these three activities where skipped.

To illustrate the kind of diagnostics one would expect, we refer to Fig. 21.11. These are diagnostics returned by ProM given an event log where we modified seven cases in such a way that all seven are noncompliant. The red arcs show the deviations. Activity *place order* is skipped three times. Activity *receive goods* is skipped four times. Using ProM one can easily drill down on the deviations. The red arcs separate the deviations from the model and one can select any arc to see the corresponding nonconforming procurement orders. Figure 21.12 shows the normative BPMN model next to the DFGs generated by Disco and Celonis. Based on these DFGs it remains partly unclear what the deviations are. It is possible to identify the three cases where activity *place order* is skipped. However, based on the arcs it is impossible to see that activity *receive goods* was skipped four times. The numbers on the arcs are not revealing this.

Some of the commercial software tools have added conformance checking capabilities in recent years, e.g., Celonis supports a variant of token-based replay.

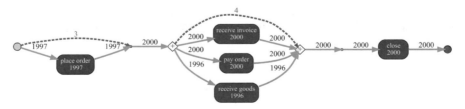

Fig. 21.11 Conformance checking applied to an event log with seven deviating cases: three cases skipped activity *place order* and four cases skipped *receive goods*

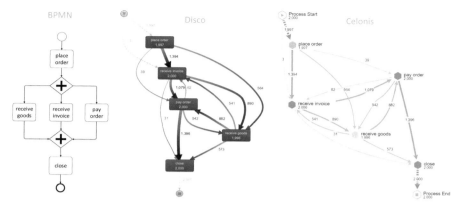

Fig. 21.12 The normative BPMN model and the DFGs generated by Disco and Celonis for the event log with seven deviating cases

However, the usability, quality of diagnostics, and scalability of existing software tools leave much to be desired.

Also from a scientific point of view, process discovery and conformance checking are still challenging. These two foundational process mining problems have not yet been solved satisfactorily and there is still a lot of room for improvement [2]. However, the current state-of-the-art techniques provide already partial solutions that perform well on real-life data sets. The hope and expectation is that commercial systems will adopt these techniques when users get more critical and expect more precise and fully correct diagnostics.

Novel Challenges

As discussed, process discovery and conformance checking are still challenging and one can expect further improvements in the coming years. However, next to these core process mining tasks, novel process mining capabilities have been identified (see Fig. 21.13). These provide new scientific and practical challenges. Here we briefly mention a few.

Fig. 21.13 Some of the challenges getting more attention in research and thus showing the anticipated development of the Process Mining discipline

- *Challenge: Bridging the gap between process modeling and process mining.* Many organizations use tools for modeling processes. With the uptake of process mining, it becomes clear that these models do not correspond to reality. Although such idealized models are valuable, the gap between discovered and hand-made models needs to be bridged [9]. A promising approach is the use of so-called hybrid process models that have a backbone formally describing the parts of the process that are clear and stable, and less rigid data-driven annotations to show the things that are less clear. Hybrid process models allow for formal reasoning, but also reveal information that cannot be captured using mainstream formal models because the behavior is too complex or there is not enough data to justify a firm conclusion [9]. Next to combining formal modeling constructs (precisely describing the possible behaviors) and informal modeling constructs (data-based annotations not allowing for any form of formal reasoning), there is also the need to deal with multiple abstraction levels. Modeled processes tend to be at a higher level of abstraction than discovered process models. One high-level activity can correspond to many low-level events at different levels of granularity. It is not easy to bridge the gap between process modeling and Process Mining. However, both need to be integrated and supported in a seamless manner. A convergence of process modeling and Process Mining tools is needed and also inevitable.
- *Challenge: Incorporating stochastic information in process models to improve conformance checking and prediction.* Frequencies of activities and process variants are essential for Process Mining. A highly frequent process variant (i.e., many cases having the same trace) should be incorporated in the corresponding process model. This is less important for a process variant that occurs only once. One can view frequencies as estimates for probabilities, e.g., if 150 cases end with activity *reject* and 50 cases end with activity *accept*, then the data suggests that there is a 75% probability of rejection and a 25% probability of acceptance. Such information is typically not incorporated in process models. For example, the normative BPMN model may have a gateway modeling the choice between activity *reject* and activity *accept*, but typically the probability is not indicated. Obviously, such information is essential for predictive Process Mining. However, the same information is vital for conformance checking. Typically, four conformance dimensions are identified: (1) *recall*: the discovered model should allow for the behavior seen in the event log (avoiding "nonfitting" behavior), (2) *precision*: the discovered model should not allow for behavior completely unrelated to what was seen in the event log (avoiding "underfitting"), (3) *generalization*: the discovered model should generalize the example behavior seen in the event log (avoiding "overfitting"), and (4) *simplicity*: the discovered model should not be unnecessarily complex. The first two are most relevant for comparing observed and modeled behavior (the other two relate more to the completeness and redundancy of event data and the understandability of the process model). Recall (often called fitness) is typically well-understood. Although there are many precision notions, it turns out to be problematic to define precision for process models without probabilities. Adding infrequent behavior to a model may significantly lower traditional precision notions. This seems counterintuitive.

Moreover, from a practical point of view, a process may be no longer be compliant if the distribution over the various paths in the model dramatically changes. These observations suggest that probabilities need to be added to process models to allow for both prediction and all forms of comparison (including conformance checking). Moreover, adding stochastics to process models also enables the interplay between Process Mining and simulation [10]. Currently, simulation is rarely used for process management. However, the uptake of process mining may lead to a revival of business process simulation.

- *Challenge: Process Mining for multiple processes using different case notions.* Traditional Process Mining techniques assume that each event refers to one case and that each case refers to one process. In reality, this is more complex [5]. There may be different intertwined processes and one event may be related to different cases (convergence) and, for a given case, there may be multiple instances of the same activity within a case (divergence). To create a traditional process model, the event data need to be "flattened." There are typically multiple choices possible, leading to different views and process models that are disconnected. Consider the classical example where the information system holds information about customer orders, products, payments, packages, and deliveries scattered over multiple tables. *Object-centric process mining* relaxes the traditional assumption that each event refers to one case [5]. An event may refer to any number of business objects and using novel process mining techniques one can discover one integrated process model showing *multiple perspectives*.

- *Challenge: Dealing with uncertain and continuous event data.* The starting point for any process mining effort is a collection of events. Normally, we assume that events are discrete and certain, i.e., we assume that each event reported has actually happened and that its attributes are correct. However, due the expanding scope of Process Mining, other types of event data are encountered. Events may be uncertain, e.g., the time may not be known exactly, the activity is not certain, and there may be multiple candidate case identifiers. Consider, e.g., sensor data that needs to be discretized or text data that needs to be preprocessed. In such settings we use classifiers that have a maximal accuracy of, e.g., 80%. By lowering thresholds we may get more "false positives" (e.g., an activity that did not really happen was added to the event log) and by increasing thresholds we get more "false negatives." The attributes of events may also have continuous variables that determine the actual meaning of the event. For example, a blood test may provide several measurements relevant to the treatment process. Another example is the event log of a wind turbine showing information about wind speeds, wind direction, voltage, etc. Such information can be used to derive an event "shut down turbine because of strong winds." Future Process Mining tools and approaches will need to be able to deal better with uncertain event data and measurements that are continuous in nature.

- *Challenge: Comparative Process Mining to identify differences between process variants over time.* Processes change over time and the same process may be performed at different locations. Hence, it is valuable to compare the corresponding process variants [11]. The relative performance often provides more insights than the absolute values. For example, what were the main differences between January and February or what are the differences between the Berlin office and the Amsterdam office? Comparing different process variants is not so easy. Compare, e.g., the DFGs in Fig. 21.5 to the DFGs in Fig. 21.8. It is not so easy to spot relevant differences. Next to changes in the process structure, also frequencies and times may change. Techniques for comparative Process Mining aim to address this [11]. This includes techniques to support the visual comparison of process variants and machine learning techniques using process-centric features [12].

- *Challenge: Causality-aware Process Mining ensuring correct and fair conclusions.* Process Mining techniques can be used to quickly uncover performance and compliance problems. Based on bottleneck analysis and conformance checking results, we can annotate event data to expose desirable and undesirable "situations" (i.e., good and bad choices, cases, routes, etc.). Moreover, using feature extraction it is possible to turn such situations into supervised learning problems and use data mining and machine learning techniques to uncover root causes for performance and compliance problems. Such a combination of techniques yields a powerful approach to automatically diagnose process-related problems. However, correlation does not imply causation. Unfortunately, these terms that are mostly misunderstood and often used interchangeably. For example, ice cream sales may strongly correlate with burglary. However, this does not imply that eating ice cream causes theft. There is a third variable (the weather) that is influencing both ice cream sales and burglary. When delays in a process correlate with deviations, it does not imply that one causes the other. Therefore, explicit causal models are required to guide root-cause analysis in Process Mining. In a causal model, relations between process features can be made explicit using a mixture of domain knowledge and statistical evidence. Similar techniques can be used to avoid unfair or even discriminating conclusions. For example, it is pointless to blame the workers that are overloaded for delays. Also the most experienced workers often take the most difficult cases possibly leading to unfair conclusions if one only considers the data.
- *Challenge: Confidentiality-aware Process Mining to avoid unintentionally leaking sensitive information.* Event data are potentially very sensitive. A few timestamped events are often enough to identify a customer or employee. Even when one removes explicit timestamps, the order of activities may already be enough for identification. Preserving confidentiality is, therefore, a primary concern in Process Mining. Current research aims at dedicated anonymization and encryption techniques. For example, one does not need to store complete cases to generate DFGs. It is sufficient to store direct succession relations without correlating all events belonging to a case. The introduction of the EU General Data Protection Regulation (GDPR) illustrates the growing importance of data privacy. Therefore, the next generation of Process Mining tools will need to support confidentiality preserving techniques. Confidentiality-aware Process Mining is part of the broader domain of *Responsible Data Science* (RDS) focusing on *fairness, accuracy, confidentiality,* and *transparency.* All four RDS

aspects are relevant for Process Mining. Not addressing these concerns may slow down the adoption of Process Mining.

The above list of challenges is far from complete. Process Mining is a relatively young, but also broad, discipline. It is interesting to compare the above list with the 11 challenges in the Process Mining Manifesto written in 2011 [13]. This shows that the field developed rapidly.

Process Hygiene

Most of the challenges mentioned in this chapter require the concerted action of Process Mining users, technology providers, and scientists. A collaborative effort is needed to make Process Mining "the new normal." Process Mining should be as normal as personal hygiene and not require a business case. Activities such as brushing your teeth, washing your hands after going to the toilet, and changing clothes do not require a business case. Process Mining can be seen as the means to ensure *Process Hygiene* (PH) or *Business Process Hygiene* (BPH). Objectively monitoring and analyzing key processes is important for the overall health and well-being of an organization. Unfortunately, managers, auditors, and accountants often still use medieval practices. Financial reporting frameworks such as the nation-specific GAAP (Generally Accepted Accounting Principles) standards still depend on the notion of materiality. As a result, sampling suffices. Given the availability of data and our ability to analyze processes, this is remarkable. Process Hygiene (PH) should not require a business case. Not using Process Mining is a sign of self-neglect showing an inability or unwillingness to manage processes properly. Hence, *not* using Process Mining should require a justification and not the other way around. Using data quality and privacy concerns as reasons to not conduct Process Mining should be considered as poor hygiene leading to "smelly processes."

References

1. van der Aalst, W.M.P.: Process Mining: Discovery, Conformance and Enhancement of Business Processes. Springer, Berlin (2011)
2. van der Aalst, W.M.P.: Process Mining: Data Science in Action. Springer, Berlin (2016)
3. Kerremans, M.: Gartner Market Guide for Process Mining. Research Note G00387812 (2019)
4. Kerremans, M.: Gartner Market Guide for Process Mining. Research Note G00353970 (2018)
5. van der Aalst, W.M.P.: Object-centric process mining: dealing with divergence and convergence in event data. In: Software Engineering and Formal Methods (SEFM 2019). Springer, Berlin (2019)
6. Koplowitz, R., et al.: Process Mining: Your Compass for Digital Transformation: The Customer Journey Is the Destination (2019)
7. Rozinat, A., van der Aalst, W.M.P.: Conformance checking of processes based on monitoring real behavior. Inf. Syst. **33**(1), 64–95 (2008)
8. Carmona, J., et al.: Conformance Checking: Relating Processes and Models. Springer, Berlin (2018)

9. van der Aalst, W.M.P., et al.: Learning hybrid process models from events: process discovery without faking confidence. In: International Conference on Business Process Management (BPM 2017). Springer, Berlin (2017)
10. van der Aalst, W.M.P.: Process mining and simulation: a match made in heaven! In: Computer Simulation Conference (SummerSim 2018). ACM Press (2018)
11. van der Aalst, W.M.P.: Process cubes: slicing, dicing, rolling up and drilling down event data for process mining. In: Asia Pacific Conference on Business Process Management (AP-BPM 2013). Springer, Berlin (2013)
12. Bolt, A., de Leoni, M., van der Aalst, W.M.P.: Process variant comparison: using event logs to detect differences in behavior and business rules. Inf. Syst. 74(1), 53–66 (2018)
13. IEEE Task Force on Process Mining: Process mining manifesto. In: Business Process Management Workshops. Springer, Berlin (2012)

Business View: Towards a Digital Enabled Organization

Lars Reinkemeyer

Abstract

The business outlook is written by the editor and considers the dimensions business expectations, potentials and benefits, technological developments, market trends, and developments of a digital workforce. The chapter has been structured on a timeline from present trends to short-, mid-, and long-term outlook, concluding in a Vision of a Digital Enabled Organization. The chapter aims to initiate thought processes, stir discussions, stimulate technical developments, and further enhance the power of Process Mining.

The following part discusses trends, predicts developments, and gives an outlook from a business perspective. It aims to initiate thought processes, stir discussions, and envisage technological developments, which can contribute to further enhance the power of Process Mining in the future.

For an outlook on Process Mining, several dimensions have to be considered: business expectations, potentials and benefits, technological developments, market trends, and developments of a digital workforce. Considering these dimensions, the following outlook has been structured on a timeline from present trends to short-, mid-, and long-term outlook, concluding in a Vision of a Digital Enabled Organization. Trends and short-term predictions are discussed based on current operational developments in the market and requirements. Mid- and long-term predictions result from numerous discussions with other Process Mining experts, process owners, and market players.

L. Reinkemeyer (✉)
University of California, Santa Barbara, Santa Barbara, CA, USA
e-mail: reinkemeyer@ucsb.edu

© Springer Nature Switzerland AG 2020
L. Reinkemeyer (ed.), *Process Mining in Action*,
https://doi.org/10.1007/978-3-030-40172-6_22

Present Trends

While Process Mining has seen some impressive development in the last couple of years, it is still a relatively "young" discipline which—as a key technology—is expected to grow progressively, to provide increasing and comprehensive business impact. The use cases presented in Part II provide a good overview of the current adoption and challenges in the operational environment. They present various cases where the technology has just recently been adopted in single departments, leaving a significant potential to expand into further functions as well as into further companies, which have not yet started to use this technology. Current usage occurs in many organizations only in single functional silos and has in many cases not yet been leveraged on a wider scale across different functions in the organization. And many other companies are still not aware of the potential of Process Mining, where this book will hopefully support to share experiences, promote awareness, and initiate interest.

Customers expect applications, which are quick to install, stable, scalable, intuitive to use, and provide an innovative vision. The current market shows a growing number of vendors and products, which differ widely in respect to price, scalability, degree of maturity, service capabilities, and portfolio. Regarding product portfolio, the main players differentiate their offerings with AI, ML, RPD, integration of Process Mining with BPM, PowerBI, offering a combined portfolio of Process Mining and RPA or with the integration of Business Intelligence tools. Most of the larger vendors have understood the necessities to drive innovation with AI and ML, to offer a comprehensive portfolio including RPA and automation either with own or with partner products, and—last but not least—to achieve customer proximity either with their own process consulting experts or by teaming up with established consulting companies.

Due to the trends for digital transformation and organizational change, digital tools such as Process Mining and AI will be of increasing relevance. But tools can only support an organizational transformation, as the following statement regarding AI underlines: "While cutting edge technology and talent are certainly needed, its equally important to align a company's culture, structure, and ways of working to support broad AI adoption. In most firms that aren't born digital, mindsets run counter to those needed for AI."[1] New digital tools and data democratization have a major influence on people, due to changing requirements and roles. Process Mining enables and supports data analysts to generate exciting new insights, but at the same time requires new skills and roles. A new generation of experts has been educated with a thorough understanding of computer science, data, and technology. In the following we will thus not only give an outlook for the Process Mining tool, but also include crucial aspects of a future digital workforce.

The mindset of a digital workforce differs significantly from the traditional mindset, e.g., regarding access to data: while the traditional approach was extremely

---

[1]Harvard Business Review 08/19.

restricted with respect to data access—with "eyes only" principles—the democratization of data is a trend which drives major change towards open data access. Access to data as well as the preparation and analytics of data was traditionally rather a task conducted by specialized experts in organizational silos. As the general perception changes towards an understanding that data is essential for today's business, data ubiquity, accessibility, and usability for everybody becomes a standard requirement.

Given the technical complexity described in the previous chapters, the future of data fabrics will focus on the capability to build analytics across different data sources and formats with an user interface which is easy and fun to use for everybody. The idea of providing data analytics in a Google-like format across an organization is, e.g., promoted by thoughtSpot.com. Similar to a Google search, analytics searches such as "predict next month's revenue" shall be answered on the fly and with tapping into different data sources. This trend is equally expected to influence Process Mining, with a strong focus on providing data in a form which is convenient to consume, assuring positive user experience.

The organizational mindsets will change towards digital affinity, understanding and using the power of data. This mindset change from "do I have to use data" towards "can I afford not to use available data" is reflected in Wil van der Aalst's term Business Process Hygiene (BPH): any organization has the responsibility to assure process efficiency, continuous improvement, and "hygiene" in respect to undesirable process deviations. Continuous hygiene can be assured with Process Mining as a mandatory prerequisite to screen actual processes, identify weak spots, deploy cleansing measures, and continuously monitor improvement progress. Similar to healthcare prevention, where hygiene is recommended and, e.g., regular dental screening has become compulsory, we imagine Process Mining to become compulsory to assure "clean" processes. Business functions will benefit in multiple forms: for audit purposes, a continuous screening based on all relevant digital traces will assure "clean" processes without violations. Support functions such as P2P or O2C will be continuously sanitized by deviations in payment terms, late payments, rework, etc., thus assuring healthy, efficient, and competitive process performance. For cross-company processes the continuous transparency provided by Process Mining can ensure hygiene in the sense of quick identification and elimination of, e.g., delays induced by the business partner. It is envisioned that the responsible process owners will need to explain in future why Process Mining has not been used to assure process hygiene.

Near Future

The operational usage of Process Mining in the last couple of years has led to some immediate business requirements, which are expected to become available in the near future:

Predictive Solutions will predict upcoming events in order to enable users to take preventative measures. It might—e.g., in procurement—be helpful to get a prediction, such as which PO will not be delivered on time. Equally, in logistics it would be helpful to get predictions for which customer orders will not be delivered on time. Based on historical data, predictions can be calculated and presented with probability thresholds: as one example, predictions can identify all supplies, which will not be received on an expended date with a probability of, e.g., 80%. Tests have been conducted for several years, e.g., based on algorithms programmed with Python on R-server, analyzing open and closed orders including times for process execution. Though some feedback has been encouraging, sufficient confidence of the predictions has not yet been reached to complement experts with domain knowledge to deploy predictions for general usage—but for sure will become available in the near future.

Proactive solutions: while big data and new tools allow unprecedented transparency, most software provides insights which require the users to search for relevant issues. Process Mining allows insights where users can identify late deliveries, rework effort, process delays, and much more. But why does the user have to search for relevant issues, spending high effort and wasting precious time while searching for relevant issues? Virtual assistants have been discussed for a long time to provide customized, individual support. In a professional environment they should support the employee by understanding tasks and daily work. Future solutions will "learn" current operations and develop skills to propose relevant exceptions proactively to the users. The software will become a smart companion, which is capable of understanding the operational process, understand exceptional issues, and propose these proactively in order to assist the user. For example, overdue payments will be presented to the user per push-mail, delayed customer deliveries will be flagged out, and potentials for automation proposed. The wake of AI and ML has made the required deep learning methods and algorithms available to screen large datasets, identify suspicious schema, and propose user-specific actions.

Usability: application development shows a strong focus on the consumer, with the requirement to provide intuitive user interfaces (UIs) which are quick to respond and fun to use. This applies to application functionalities as well as access to data. Trends move away from predefined, static dashboards which have been built centrally and were designed to provide a standard framework. Instead, dynamic sets of data analytics cubes are requested which can individually and intuitively be consumed. This convenience to access data will be complemented by natural language processing, enabling users to communicate with technology via voice input.

The *Customer Journey* has become an emerging focus topic. A majority of the participants at the first global Process Mining conference, ICPM 2019, considered this a focus area for Process Mining. Customer centricity and customer obsession are strategic priorities and require a thorough understanding and continuous optimization of customer interactions. Imagine the benefit of a full transparency of the complete customer journey, based on digital traces generated from customer interactions on webpages, mobile apps, partner sites, purchasing portals, customer interaction center, order placement, deliveries, payments, etc. Insights will allow to combine customer journey maps with actual customer interactions. Customer satisfaction will be significantly increased by understanding actual customer behavior, predicting order delays, and allowing to proactively serve customer demands. Single use cases have shown initial improvements for customer interaction (e.g., Uber, EDP), but this is only the beginning and the heterogeneity of multiple data sources causes a challenge for achieving full transparency. Purpose has been defined with multiple ideas for valuable Process Mining use cases and the challenge of collecting Processtraces from various sources is expected to be resolved in the near future.

Data collection and preparation: while projects today still require major effort to identify, collect, and prepare event logs, there are promising first approaches to automate these tasks. In particular structured data from homogeneous systems (e.g., SAP ERP) will be identified and read by standard algorithms. Preparing, e.g., a P2P reporting across multiple systems around the world will become possible with much less customizing effort in the near future. Automated discovery of event logs and process data is expected to become possible building on the growing experience gained from data preparation and technical innovations. Machine learning algorithms will understand the format and structure of data in similar source systems, facilitating an automation of data collection and preparation. Complementary to established ERP platforms, workflow platforms such as Pegasystems, Salesforce, and ServiceNow play an increasing role for process management and will come into scope. Data collection and preparation across these different types of platforms will become available to combine, e.g., financial and customer data.

While Process Mining is based on event logs from backend systems, *Robotic Process Discovery* (RPD) allows for process insights based on recorded user activities. Captured activities are mouse clicks, keystrokes, application inputs, and field entries, thus providing a much deeper understanding of individual working behavior. This approach—also referred to as Task Mining—allows to learn actual human activity with the purpose of identifying potentials for improvement. Any activity can be recorded, including activities in Outlook or Excel, where no log files are available and data is stored in unstructured format. While RPD provides a micro-picture of individual behavior and thus allows optimization of individual tasks, it does not allow insights into overarching operational processes, which can only be visualized with Process Mining. RPD can complement Process Mining as a "magnified" analysis of actual behavior and some Process Mining vendors have started to invest into this field. Solutions are expected to mature quickly and become a valuable support for operational experts.

Technology will evolve using new combinations of natural language processing, computer vision, sequence modeling, anomaly detection, and ML. Like many other innovations, these are hot topics which get heavily promoted as eye catchers on social media, fairs, and conferences. While first solutions are offered on the market, a general user acceptance has not yet been achieved as users with years of domain-specific experience request high accuracy to accept virtual assistants. Unlike in the B2C arena, where AI has been widely adopted, the professional business environment turns out to be more challenging as it requires process-specific domain know-how. In the backend, technological development will need to cope with an ever-increasing data volume.

Cloud technology: Storing digital traces in a public cloud has become commonly accepted and will support the possibilities to use proven algorithm for extraction and customization of data, deploy standard use cases, and benefit from analytics available in the cloud. Hosted AI is expected to become attractive and available in the form of Software as a Service (SaaS) and accessible with standardized Application Programming Interfaces (APIs) to provide applications, technology, and best practices to a wide number of users.

Platform Solutions: Digital technologies evolve exponentially and have led to a significant shift of competitive advantages towards platforms. Many markets have been disrupted by platforms, and Process Mining appears a suitable technology which can be further expanded via platforms. In the book "Platform Revolution,"[2] G. Parker et al., describe how platforms can leverage network effects and scale more efficiently, with many aspects appearing easily adoptable for Process Mining. A platform can be attractive for software vendors, app developers, customers, and consultants: vendors will be offered a channel to sell software as a service. With the increasing number of applications which have proven valuable, app developers will provide dedicated applications and launch standardized use cases on platforms. Use cases such as P2P or O2C will be offered in the form of apps for individual subscription. Customers will have the freedom to choose ready-to-use apps and software from different vendors. Apps will be available with software from different Process Mining vendors in order to allow the consumer to choose the most suitable product. Data security, which remains a major challenge, will be assured with encapsulated data objects, which are only readable by the data owner or authorized consumers. Consulting companies see the value of Process Mining, which complements their domain and process expertise, as an exciting field for growth. Besides providing infrastructure, technology stack and standard apps, the consulting company will bring in knowledge regarding different functions and processes, gained during many years of process consultancy projects.

IIoT Platforms: The Industrial Internet of Things (IIoT) has set the technical foundation for an extensive access to event logs, as devices become digitally connected to an internet-hosted platform, thus allowing easier access to digital footprints, which are generated from these devices. IIoT platforms such as MindSphere already today receive data from millions of single devices, including relevant event logs. Purpose can, e.g., be defined to understand manufacturing processes based on the event logs from multiple machines—even across machines at different sites. The collection of event logs from different machines, sites, and companies will allow new use cases such as the visualization of cross-company supply chain processes or intercompany benchmarking. As a crucial benefit, the IIoT platforms provide a standardized and secured environment and protocol, which has been adopted to industrial requirements.

Real-time transparency: Most of today's reports are updated on a daily basis, despite a demand for real-time insights. Optimizing shop-floor manufacturing

---

[2]Recommended reading: "Platform Revolution" from G. Parker, M. Alstyne, S. Choudary (2016).

processes with a real-time insight will, e.g., allow immediate decision for optimum routing of parts in different manufacturing lines. Customer interaction tracking will allow immediate insights for quick reaction and timely interaction with customers. While the traditional onsite infrastructure typically not yet supports a real-time reporting for large amounts of data, it is anticipated that cloud solutions, streamed event data, and analytics applied directly on the source data will allow real-time transparency.

Skills: Like in healthcare, where innovative technology requires new skills and specialists, there is an increasing demand for specialists in process analytics. With a thorough understanding of process management as well as data analytics, these specialists will develop the skills to bridge the gap to the domain know-how from functional departments and apply new digital tools. In a paradigm shift, focus moves towards technology-based services. The strategic workforce planning needs to consider these requirements and facilitate the qualification of employees for future tasks and necessary skills with appropriate training and upskilling. Numerous research has shown that this shift in qualifications will lead to new roles and an overall net increase in jobs.

Digital Twin of an Organization: DTOs—as described in Chap. 7—allow to visualize and understand actual processes. Near-future solutions will provide DTOs which allow dynamic simulations of processes based on actual data, thus supporting assessments of optimization potentials. Scenarios will become possible by simulating process improvements and calculating saving potentials for different process designs. Factors like process redesign or deployment of RPA bots will be simulated to assess impact. In the future, DTOs could be integrated with causal deep learning techniques allowing to understand why and how change to a process would cause certain effects, which further enhances the simulation potential. During the implementation phase of newly designed and simulated processes, DTOs will allow continuous monitoring and support early adjustments whenever deviations from the simulations occur.

Midterm Future

Self-learning and self-optimizing systems: With AI becoming more mature and suitable to assist even in environments where profound high domain knowledge is required, technology will evolve towards self-learning and self-optimization. Imagine a process execution system which is autonomously capable to learn, i.e., to detect and resolve process inefficiencies. Similar to self-driving cars, there will be "self-driving" Process Mining tools which are capable to learn factors which determine efficient process flows and autonomously initiate measures to optimize process efficiency, including optimization of variants and reduction of process exceptions.

While the impact of *AI*, which has been experienced in today's operational use cases has been limited, it will grow up to its promises. Some innovative providers show exciting use cases with virtual process analysts discovering and documenting actual processes by imitation learning. A virtual digital companion learns from actual and optimum process handling and is thus trained to become an accepted artificial coworker, understanding also complex domain know-how, which is the big challenge in the B2B environment. Virtual companions are trained to identify and remediate process flaws, which can start with simple, repeatable process tasks, such as removal of delivery blocks. Besides all excitement about AI, it must remain explainable in order to ensure ethical data usage with clear transparency about what and how AI is applied. AI governance will play an increasing role and will have a significant impact on the acceptance of these new technologies in a corporate environment.

Benchmarking: Process Mining makes process efficiency measurable and transparent. As it is based on big data and facts, it is predetermined for benchmarking purposes. Standard processes such as P2P and O2C will be benchmarked on operational performances, such as automation rate, throughput time, or rework. With digital traces available on Platforms and in the cloud this will also become available as a self-service, where companies can access benchmark data—based on appropriate data anonymization—to assess their own performance. And consulting companies will be able to lift cross-company benchmarking analysis to a new level of data foundation, as benchmarking can be conducted based on the full set of all relevant events from different players.

Overall market: The synergies of today's separate categories Process Mining, RPA, RPD, and technologies, such as AI and ML, will lead to further mergers and acquisitions between market players. Providers will strive to provide a concise portfolio to support the digital transformation of companies. We expect intelligent process management suites to evolve in order to optimize business process efficiency with a selection of different digital tools. This will not only be supported by acquisitions, but also by strong partnerships between vendors from the different categories in order to develop joint solutions. Traditional providers from other categories such as ERP, CRM, or workflow management tools are expected to appreciate the benefits of this key technology and strive to extend their portfolio either by partnerships or acquisitions.

Long-Term Future

Intercompany: The long-term perspective provides significant economic and ecologic benefits through optimization of cross-company supply chains, based on data from different companies and sources. Process optimization will become possible for intercompany value chains, from supplier, manufacturer, freight forwarder to customer. Companies like Slync already today offer multiparty supply chain interaction with a high degree of automation, across different organizations and data sources. The value proposition offers logistics orchestration across manufacturers, suppliers, freight forwarder, and customers. With Process Mining, this could be taken to a new level by understanding the extended e2e process chains. On-time delivery, integrated manufacturing, and optimization of stock/working capital are just a few benefits of a transparent supply chain process, which can be monitored and managed with the support of Process Mining. Several of the use cases which were presented in Part II with an intracompany focus can be extended to intercompany, for integrated process optimization. Empowering the business partner with access to own process data will allow both parties to benefit. Besides economic benefits, this will lead to a sustainable ecological optimization due to the wholistic approach which will allow to reduce, e.g., the number of empty deliveries, reduce waste and allow for a better resource management and more sustainable business.

A *Sustainability Revolution* can be supported by technological innovations such as Process Mining, but also affects everybody. Think about process inefficiencies in your immediate environment and how better process efficiency could support sustainability: from traffic congestions to waiting times in hospitals, from wasted times in call center queues to waiting times for bureaucratic decisions, from delayed goods deliveries to delayed flight arrival. Process inefficiencies are omnipresent, producing friction, waste, and avoidable emission. Understanding the e2e processes allows to track down inefficiencies and reduce waste in time and resources. While SCM is probably the primary field, where Process Mining can support a sustainability revolution, CRM and other functions can equally support as ecological driver. The management of resources in ERP systems (financials, materials, assets and HR) will become more efficient with Process Mining, thus allowing to optimize scarce resources. In a world with more than 7.5 billion people and increasing issues due to limited resources, this may become a strong purpose.

B2C: While the primary focus on Process Mining to date has been on business-to-business (B2B) processes, there is a huge potential for process optimization in the business-to-consumer (B2C) field. Understanding consumer interactions with the additional dimensions of time and sequence allows to better interpret and predict consumer behavior. B2C use cases could include, e.g., activity tracking for the timing and sequence of user clicks on shopping pages or in social media platforms. Understanding strategies, how users approach challenges such as search for restaurants or music, appear valuable and might allow for trail prediction. As another example, the insight into the search sequence for web offerings could—based on large amounts of activities—not only be interesting for psychometric analysis, but also for product management and sales.

Vision of a Digital Enabled Organization

Imagine an organization which has been automated for a majority of standard processes, such as procurement of indirect material, financial transactions, order deliveries, and customer order processing. Standard tasks are conducted automatically by an AI, which is capable to learn not only how to execute standard cases, but also minor exceptions, conducting immediate actions and corrections. This "intelligent system" processes the large majority of all activities with zero human touch, and humans only interfere exception based, thus providing a high process reliability at minimum transactional cost.

Data ingestion from diverse source systems is supported by AI, which allows to identify and customize structured and unstructured data from various sources such as ERP or workflow systems. Cloud technology is commonly established as a basis for data hosting, collaboration, and data mining, with the application providers applying continuous monitoring and optimization. Streamed event data allows real-time process analytics for immediate reaction, e.g., for customer interaction. Platforms offer standard apps for process execution in a secure environment and share best practices for process handling and monitoring.

As the majority of operational processes have been fully automated, the focus of Process Mining changes. There will be less demand for transparency and insight with respect to today's focus areas. Standard support processes such as P2P and O2C provide decreasing marginal benefits as they are widely optimized and the focus shifts towards more challenging processes, such as customer interaction, manufacturing, HR, and legal. Besides intercompany automation, cross-organization process optimization is happening in integrated supply chain process flows. Exception-based activities remain in focus, as they require optimization with appropriate digital tools. Similar to tele-medicine, remote diagnosis and optimization of processes and automation will be available through dedicated Process Mining Analysts, who are alerted by intelligent virtual assistants, which conduct a continuous real-time monitoring and provide predictive and proactive alerting.

The role of humans has changed significantly: mundane tasks have been completely automated and new tasks and roles emerged instead. The focus of human responsibility has changed towards data analytics and steering, using tools which are provided by the digital enabled organization. Process analysts use digital tools such as virtual assistants, which collect data from Process Mining and RPD, supported by DTOs, thus empowering the digital enabled organization. Value generation shifts towards service innovation. In their book *Dreams and Details*,[3] Snabe and Trolle describe how to reinvent business from a position of strength and with a compelling vision. An innovative "Digital Enabled Organization" could provide the dream to set the mindset and framework to unleash the human and digital potential.

As a positive ecological contribution, process optimization has yielded significant reduction in carbon footprint, e.g., due to reduction in empty trips and optimization of routings. Transactional costs have been reduced to a minimum.

It remains to be seen how this brave new world develops.

---

[3]Dream and Details, Jim Hageman Snabe and Mikael Trolle, 2019.

About the Editor

Dr. Lars Reinkemeyer is a visiting scholar at the University of California, Santa Barbara, and senior executive of Siemens AG. Since 2014 he leveraged Process Mining technology in close collaboration with Siemens' functional departments like sales, logistics, procurement, accounting and has established a global community in excess of 6000 active users around the world, supporting the company's digital transformation. As head of the Global Process Mining Services at Siemens Corporate IT, he has built a team of experts located in Germany, Portugal, and India providing cross-functional Process Mining services.

Dr. Reinkemeyer joined Siemens AG in 1994, right after he earned a master's degree in Business Administration and a PhD from the University of Cologne, from which he graduated summa cum laude. He joined Siemens as Product and Regional Manager and was delegated to Siemens Australia as International Account Manager in 1996. In 1998, he signed on as General Manager at Oztrak Europe GmbH, gaining some hands-on start-up experiences. In 2000, he joined Atoss Software AG as a Director of International Sales. Dr. Reinkemeyer rejoined Siemens AG in 2001, where he held various international leadership positions in strategy, compliance, and IT. He is a guest speaker at Stanford Graduate School of Business and regular speaker on international conferences.

© Springer Nature Switzerland AG 2020   207
L. Reinkemeyer (ed.), *Process Mining in Action*,
https://doi.org/10.1007/978-3-030-40172-6

Printed in the United States
by Baker & Taylor Publisher Services